To Roy

With best wishes

Warren L

10. 11. 2009

A SILENT GENE THEORY OF EVOLUTION

Warwick Collins

The University of Buckingham Press

First published in 2009 by
The University of
Buckingham Press
Yeomanry House, Hunter Street
Buckingham MK18 1EG
United Kingdom

Second impression

ISBN 978-0-9554642-8-7

Printed by the MPG Books Group in the UK

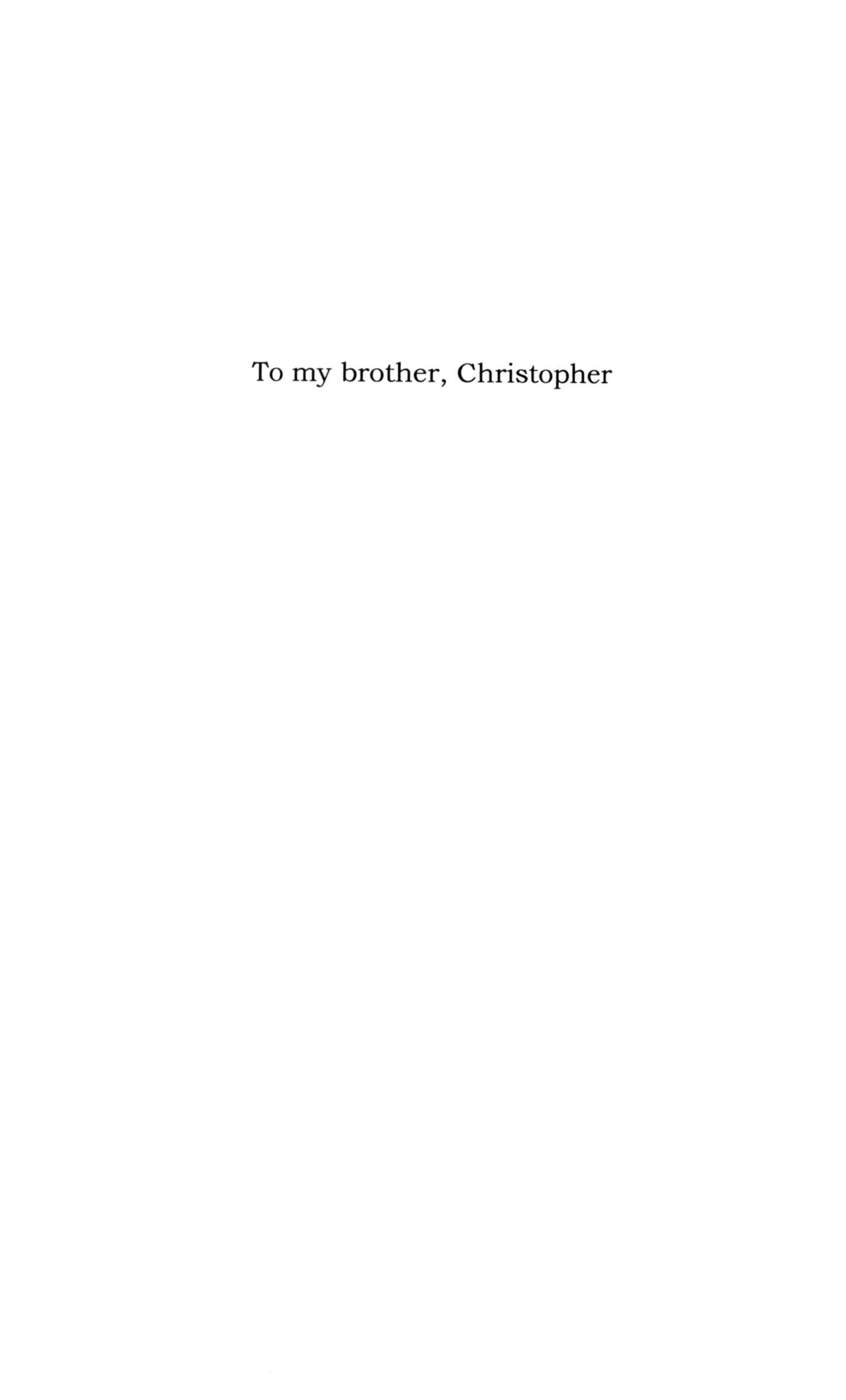

To my brother, Christopher

TABLE OF CONTENTS

FOREWORD

Donald W Braben

Visiting Professor, Department of Geological Sciences
University College, London

The rise in the field of molecular biology, and the discovery of the structure of the molecule of life — DNA — by Crick and Watson, in 1953, together with a vast number of other discoveries, have transformed our understanding of genetics. We now know that life is specified by long, connected sequences of molecules called nucleotides. These molecular types, the same for all species from ants to zebras, come in four distinct varieties. They constitute the genetic code and the blueprint for life.

The Human Genome project, launched less than twenty years ago, has now succeeded in meeting its goal of identifying the entire genetic sequence of a human being. The sequences of other species — including viruses, bacteria and fungi — have also been completed.

Scientists are now awash with data. Merely writing down the linear sequence of the human genome would fill a ten million page book. Mountains of facts, however, are not the same as understanding. We still grasp very little, and in the rush to publish and protect their careers, scientists are often denied the time for quiet contemplation.

Despite our progress, important areas remain to be understood. One of the most profound mysteries in biology is that the genetic sequences coding for the more complex forms of life — plants and animals — are only a small part of the whole genome. For humans, it is only about 1.5%. Thus some 98.5% of the human genome has no known function. Until recently, the latter was often called "junk" — the clear implication being that this vast array of inert genes has no significant function. Yet every time a living cell divides, the entire genome — junk and all — is faithfully duplicated.

This central mystery is at last beginning to attract the attention it deserves. A number of years ago, Warwick Collins came to realise that our current understanding of evolution could not explain how life survives radical environmental change. Before selection can occur, there has to be a range of variation within a population on which adaptation can act. Yet in adapting a population, natural selection reduces variation in favour of an evolutionary optimum for that given environment. If natural selection actively consumes variation in the process of adaptation, Collins asked, what is the true source of the continuously replenished variation throughout evolutionary systems? And if it is an identifiable process, how precisely does it overcome the continuous variation-reducing effects of natural selection?

Collins has made a good *prima facie* case for looking at the vast unexplained sections of the genome in a new way. It is a theory, of course, and will need to be tested rigorously. But new thinking is notoriously difficult to get on the agenda nowadays. Dogma and consensus dominate the governance of science. A new idea must eventually satisfy the strictest empirical criteria but, more immediately, it has to compete with the power and influence of those individuals who hold contrary views. Collins' story is fascinating in this respect too, and it is far from being untypical.

To resolve this central problem, Collins proposed that large quantities of inert or "silent" genes are necessary to give organisms the variability to adapt. According to his argument, the fact that the silent genes do not "code" for physical characteristics (as far as we know) means that they are not subject to the normal processes of natural selection acting on the individual organism or phenotype, and so they can mutate and evolve freely over long periods of time without adversely affecting the host organisms.

To reach this conclusion, Collins clearly explains his audacious theory. He deploys one of the greatest mysteries in modern genetics — what is the function of the vast number of genes in the genome which are silent? — to resolve one of the greatest problems in evolutionary theory: what is the source of variation?

In the course of setting out his case, Collins makes the additional argument that, if variation must be present *before* natural selection can act, a strong theory of variation has a more powerful claim to be a general theory of evolution than a strong theory of adaptation. Thus Collins is proposing a general evolutionary theory which — if it continues to be supported by the data — may in due course come to rival Darwin's theory that evolution is driven by natural selection.

INTRODUCTION

No doubt many of us are subject to obsessions. My own is that evolutionary theory is too important to be left entirely to a small group of established professional biologists with a strong interest in the *status quo.* Evolution by natural means has become the central secular explanation for our own origins and that of life in general. More than any form of religion, or even the abstract notion of a single presiding deity, it is now the primary explanation of our common origination. It is the most important strand of our secular heritage, and it belongs, in this sense, to us all.

We should be careful too that Charles Darwin's remarkable and varied legacy does not fall into the exclusive ownership of an even smaller group of evolutionary ideologues who persist in propounding a narrow and doctrinaire interpretation of his work. He is much too large a figure to be so constrained. It is an interesting exercise to consider what Darwin himself might think of the modern keepers of his flame who, when they are not fighting bitter wars with one another on relatively minor differences of textual interpretation and denouncing one another with all the fervour of religious zealots, are happy to assert to a wider public that Darwin's theory of natural selection is an absolute and unassailable truth.

We know, not least by exercise of common sense, that this latter assertion should be treated on its merits. Given Darwin's limited knowledge of the mechanism of inheritance (he knew nothing of genes, for example) it is worth repeating that the theory that evolution is driven by natural selection is not the word of an omnipotent deity, but the best explanation of evolution by natural processes which a brilliant nineteenth-century naturalist could offer within the constraints of his time and place.

Consequently, when I speak of "Darwinism" or the "Darwinians" in what follows, I am not referring to Darwin's unprecedentedly rich and varied contribution to evolutionary theory and natural history, but to the academic biological establishment who claim to speak in his name. Marx famously and wittily said that he was not a Marxist. Perhaps we should leave open the question of whether Darwin, were he alive to today, would consider himself a Darwinist.

Turning to my own particular interest in evolutionary theory, I am a novelist by occupation, though at the time of writing my work is probably better known in translation in various European languages than in my own. I have been interested in evolutionary theory since I was a child, but my only brush with formal education in science was as a student of biology at the University of Sussex in the mid 1970s, where my tutor was the leading theoretical biologist Professor John Maynard Smith.

For what it is worth, until I found myself at Sussex, studying biology, I had not intended to involve myself with science in any formal manner, and my brief association with biology came about almost accidentally. In my early twenties, and largely for my own interest, I wrote a paper on evolutionary theory addressing so-called "secondary sexual characteristics" such as the tail feathers of the male pheasant — structures which are anomalous in terms of Darwinian natural selection because they appear to confer a survival disadvantage on their owner. Somewhat speculatively, I sent my paper to the biological journal *Genetics*.

Science journals are subject to what is called "peer review", which is to say that the submitted paper is adjudicated or refereed by a leading expert in the field. John Thoday, then professor of genetics at Cambridge, refereed my paper and subsequently invited me to do research work. I was flattered, but I told him that unfortunately I had no scientific qualifications. He tutted awhile, and said if I wished to pursue my researches at a later date I could either undertake a degree in Natural Science Tripos at Cambridge, which at that time would have involved a considerable amount of inorganic chemistry and some physics — subjects not closely connected to my main interest — or a "pure" zoology course somewhere else. I ended up at Sussex University, where the distinguished evolutionary theorist John Maynard Smith was my tutor.

Maynard Smith treated me very well, and asked me to join a group of postgraduates and younger faculty members for regular discussions about theoretical biology. One evening, after one of our free-wheeling debates on various aspects of evolutionary theory, Maynard Smith and I adjourned to have a drink at a nearby pub. During our discussion, I mentioned that I had severe doubts about the Darwinian theory that natural selection of individuals drives evolution. Maynard Smith asked me to expound my views and, with a certain amount of trepidation, I set out the following points.

It seemed to me that a grand theory of evolution should explain, with some precision, certain salient features of the evolutionary landscape. What bothered me was this. Evolution is the story of simple, generalised organisms developing over time into more variable, complex organisms. Unfortunately, the natural selection of individuals cannot explain this, because wherever you examine its action in detail, in any specific population, it acts *against* variation in favour of optimisation for a given environment. If there is a range of variants, A, B, C, D, in a given environment, one of those variants is likely to have a slight advantage, and the others are likely to be reduced or eliminated over time. Always and everywhere, it seemed to me, the direction of natural selection is *counter-variation*.

I told Maynard Smith I thought natural selection existed, was powerful, and was entirely worth studying, but that if it were the predominant force in evolution, it would drive the process backwards towards more simple, generalised organisms. It would do so by constantly thinning out variation in any population of complex organisms until only one optimum type survived in any given environment. Since the environment is dynamic (for example, in the form of the diseases and parasites which assail any given population), when the environment changed radically there would not be sufficient variation within the population to adjust or evolve. In such circumstances, the likelihood was that in due course the population would die out. If this were so, then natural selection offered an elegant theory of natural extinction, but not (as Darwin claimed) of the origin of new species.

If the overall effects of the process were to be considered, natural selection would eliminate the most complex, variant species in favour of simple, rapidly replicating generalists. It would thus tend to reverse the main process of evolution, driving it backwards towards more simple but numerous single-celled organisms.

Maynard Smith asked me why I did not accept the standard view that variation is caused by random mutations in the genes which code for physical characteristics. I responded by suggesting that (as we shall see below) there is very little empirical evidence for the type of mutation of coding genes which would be necessary to explain variation in Darwinian terms. In addition, there are good theoretical reasons why there should be such a lack of empirical evidence (this will be discussed below in some detail). And even if such evidence could be found, the quantities of variation required would not be remotely sufficient to counteract the continuous variation-reducing effects of natural selection throughout evolutionary systems.

Most important, from my perspective, was the problem of "intermediate" mutations. The difference between one effective coding gene and its closest relation was seldom one mutation but more often than not tens or hundreds or thousands of mutations. The likelihood that each of these mutations would all occur at once to create a new, effective coding gene was vanishingly small. The individual intermediate mutations were far more likely to have accumulated gradually over time. But this led to another crucial problem when considering coding genes. Each of the intermediate mutations which affected the host organism was likely to be deleterious to the host organism (since the majority of mutations are deleterious) and was therefore likely to be selected out by natural selection acting on the host organisms.

Taking all these factors into account, I believed there must be another natural process, capable of generating variation, which was also sufficiently powerful to offset the continuous variation-reducing effects of natural selection.

At the time of my brief stay at Sussex University, Maynard Smith was one of the world's leading theorists within the field of evolutionary biology, and quite reasonably enough these doubts of mine did not go down well with him. He asked me whether I could identify another countervailing process which would offset the variation-reducing effects of natural selection, and I said that I could not. Maynard Smith then issued a warning. He said that if I did not accept the Darwinian theory that evolution is driven by the natural selection of individuals, it would be difficult for me to prosper in biology. He thought I had a promising career if I worked within the framework of natural selection, but if not my views would isolate me.

I replied that I thought the essence of science was constantly to question the truth of theories, not least major theories. He responded that, on the contrary, science is a collaborative community which functions through shared beliefs. In hindsight, I happen to believe that his *de facto* description of science is probably closer to the way that science actually works than my own naïve belief that science should seek out the truth at all costs. To cut a long story short, I left biology soon afterwards and made my eventual career as a novelist. It has been a somewhat peripatetic existence, but it is a decision I have never regretted.

For the record, I have always regarded Maynard Smith as a superb scientist and theorist. Furthermore, in any discipline agreed frameworks are necessary for sustained advance, and some degree of enforcement is necessary against anarchists and inveterate trouble-makers such as myself. Those happened to be the roles in which our particular circumstances cast us. If the theory I will now outline, which proposes that "silent genes" are the main cause of complex evolution, ever becomes more widely accepted, it will be at least partly due to Maynard Smith's inspired teaching, and his clarification of certain key evolutionary issues.

Meanwhile, having left biology, in my spare time I continued to think about an evolutionary process which would explain variation. During that time I made no real progress on the problem, but I persevered in thinking about it, and obsessively discussed the subject with anyone foolish enough to listen to me. Some twenty-five years later, in March 2000, I began a scientific paper called *A Silent Gene Theory of Evolution.*

As mentioned above, I had wondered whether there was a process, preferably within the genes themselves, which could generate variation on a substantial scale, and independently of the process of natural selection. The chief difficulties continued to appear insurmountable. One of the functions of genes is to specify or "code for" physical characteristics. The main problem is that coding genes are likely to be eliminated when natural selection acts against the physical variation for which they code. For example, if blue eyes are mildly disadvantageous compared with brown eyes in the harsh light of the tropics, then in tropical environments individuals with blue eyes will be selected against over time. In the process, through the slow relative elimination of blue-eyed individuals, the blue-eyed genes will be proportionately reduced in the gene pool. This in turn leads to an important general principle. *Genes which code for physical characteristics are unlikely to be able, as a class, to resist the variation-depleting effects of natural selection acting on physical characters, because the coding genes themselves are effectively subject to that same natural selective process.*

In order to find purchase in this problem, I began to consider whether there might be another class of genes, which I called "silent" genes. These entities (in my own mind at the time, entirely theoretical) would not code for physical characteristics, but would exist "inactive" in significant numbers within the genome. When the individual organism reproduced they would be replicated alongside the other genes. Precisely because they were inactive or inert, they could change and mutate gradually over long stretches of time outside the constant counter-variation effects of natural selection on the coding genes. In due course, probably through random processes, these non-coding genes could begin to code at a later stage, introducing new variation, which would then be sifted by natural selection.

Perhaps the reader will agree that this is a somewhat exotic theory. Because I was interested in Maynard Smith's reaction, in March 2000 I wrote to my former tutor, enquiring whether he remembered me, and asking if he might care to glance at my initial speculative paper.

Given that we had parted some twenty-five years before after a sincere disagreement between us, and we had not been in contact since, I expected a flea in my ear or perhaps, at the very least, an eloquent silence. But, as it turned out, Maynard Smith was kindness personified. He wrote that he did indeed remember me, and that he found my paper interesting. He observed that, from the properties I described, "silent" genes were obviously "junk" genes.

This was a remarkable assertion. Prior to writing my highly speculative theoretical paper, I had little reason to investigate junk genes, and the notion that these physical entities corresponded to my theoretical description of silent genes was new to me. I admit that at the time I had very little knowledge of the junk genes. If I had been asked what junk genes were, I probably would have answered that they were malfunctioning coding genes. I make this point to stress that at the time at which I wrote to Maynard Smith enquiring whether he knew of any genes which corresponded to my description, the concept of "inert" genes was, for me at least, entirely a product of my extremely strange and tentative theory.

Apart from being inactive or inert, it appeared that "junk" genes are also present in the huge numbers I thought would be necessary to generate significant variation. In humans, for example, only about 1.5% of genes in the genome are active or coding genes; an astonishing 98.5% approximately of all human genes are silent or "junk" genes.

Junk genes are created mainly by random replication of other genes. The resulting "junkyard" of slowly mutating genes exists within the genome alongside the coding genes, and is reproduced faithfully along with the coding genes when the individual reproduces. But why were there so many? Why were there more than fifty times as many silent or junk genes as active or coding genes in the human genome? And what, if anything, was their significance?

In terms of established theory, junk genes have always been an anomaly, as their very name implies. According to standard Darwinian theory, genes code for physical characteristics and natural selection acts on these physical characteristics, so that the ratios of genes in the population rise or fall according to the benefits they confer on the individual. This is the classical selective process of evolution, and junk genes appear to play no part in this scheme.

Having made the connection between my theoretical "silent" genes and junk genes, Maynard Smith nevertheless strongly disagreed with my thesis that junk genes were responsible for generating variation. In my brief paper I had made the point that in order to generate physical variation, junk genes would have to become coding genes at some stage. Maynard Smith responded that to his knowledge there were no known examples of junk genes becoming coding genes. As we will see later, there have been remarkable new discoveries in this field, which completely reverse this assertion. But at the time, my former tutor had made a perfectly valid point.

Meanwhile, Maynard Smith made the kind offer that I should send my paper to the editors of the *Journal of Theoretical Biology* with his recommendation that they consider it for publication. Despite this generous offer of an introduction, given the system of peer review, I had very little faith in the outcome. It seemed to me that when a leading biologist encountered my apparently heretical paper, he (or she) would be unlikely to advocate publication. I was willing, however, to submit the paper in the hope that my doubts would be proved wrong.

The paper was submitted twice, and on both occasions the same referee rejected it for publication. At the first submission, the referee thought that the silent gene theory was a reworking of a theory of evolution by the great Japanese geneticist Motoo Kimura, called "neutral evolution". Writing a response to the referee's first critique, I was able to persuade the editorship of the journal that my theory had nothing to do with Kimura's neutral theory, and that the referee was simply mistaken. My paper was then sent to the same referee again. On this occasion the referee attacked my theory from a wholly different perspective. Having failed to identify silent gene theory with Kimura's theory of neutral evolution, this time he proposed that the silent gene theory of evolution was based on group selection, a major heresy within orthodox Darwinism. But almost as an aside, as though to chide me for my lack of knowledge, during his critique he made the remark that there were at least four known examples of junk genes becoming coding genes.

This was an astonishing discovery, because it provided the final missing link in the theoretical outline of a silent gene, and overturned the principal objection to my theory which had been put forward in good faith by Maynard Smith.

If I may summarise the story to date, I had begun doubting whether Darwinian natural selection at the individual level can explain variation, not least because natural selection, acting unchecked, constantly reduces variation in any population or species. In order to provide a separate theoretical source of variation, in March 2000 I had posited the exotic theory of silent genes. My former tutor Maynard Smith had proposed that my description of silent genes corresponded to the junk genes, which existed in enormous and unexplained quantities in the genome. But he raised a key objection. If my theory was correct, the part that was missing from my description of silent genes was any empirical evidence that silent or junk genes could eventually become coding genes. Now the referee had casually informed me there were no less than four known and identified examples of junk genes becoming coding genes.

Those who are stubborn or foolish enough to propose a new theory sometimes experience an eerie sense of premonition when a piece of evidence falls their way. On two occasions now — when Maynard Smith confirmed the existence of huge numbers of silent genes as predicted by my theory, and when the referee had mentioned that silent or junk genes can become coding genes — I had felt a similar strange sensation that my reasoning was on the right track.

Meanwhile, I continued to think about my silent gene thesis and worked on it intermittently between other writing work. Further reports emerged in the literature of research indicating the increasingly complex contributions of the great majority of genes in the eukaryote genome which are silent.

In 2004 I thought of returning to my former tutor in the hope of further debate on the subject of the constantly expanding research evidence that silent genes could become coding genes. But before I could do so I learnt that Maynard Smith had died on April 19, at the age of 84. It was difficult to imagine theoretical biology without his powerful presence at its centre. The obituaries all attested to his inspiring example to younger biologists, his cheerfulness, humour, and amenability. I agree wholeheartedly with those views. He was in many respects unique, and he leaves behind him a singular legacy in evolutionary biology.

The empirical weakness of natural selection

I submit that the finding that junk genes can form coding genes is even more significant when placed in context. Mutations, or changes in the genes, may be either point mutations, involving minor changes in the genetic material (often single base-pair substitutions, which tend to have little or no effect) or macro-mutations, involving for example significant deletions or additions of genetic material. Most macro-mutations are deleterious to the organism, and many are lethal. For obvious reasons, what interests evolutionary theorists are those macro-mutations involving significant changes in the genetic material which in turn lead to new and useful physical structures. These are sometimes called benign macro-mutations. In what follows, the term "mutation" will also refer to macro-mutations.

One of the great weaknesses in neo-Darwinian theory is the belief that new variation is created by mutations in the coding genes. The problem is that no one so far has been able to point to a single clear example of a coding gene mutating directly into another coding gene without an intermediate silent phase. Because of this weakness, Motoo Kimura, the great Japanese geneticist, has described the neo-Darwinian theory of mutation as largely "mythic".

When one compares the two sources of mutation — the coding genes and the silent or junk genes — one perhaps begins to see why this is so. For a coding gene to become a new mutational gene would typically require not one but a whole series of random changes, involving probable deletions of material, probable additions, not to speak of numerous and cumulative point changes. If, however, the gene in question is a coding gene, then a number of these intermediate stages are likely to be detrimental to the organism (on the principle that the great majority of mutations are detrimental, and only a small minority are beneficial). So the intermediate states will be eliminated over time through natural selection acting on the deleterious physical characters for which they code.

In silent genes, by contrast, because the genes do not code for physical characters, they are immune from natural selection. Natural selection, acting on the physical characteristics of the organism (collectively called the phenotype) cannot "see" the silent genes, and therefore the silent genes can pass through an almost infinite number of intermediate mutational stages over long stretches of time *without deleterious effect on the host organisms concerned.*

When assessing whether new variant mutations are likely to arise from the coding genes or the silent genes, we should not only take into account that there are usually more silent genes than coding genes in the eukaryote genome — the genome which is characteristic of all complex and multicellular organisms. We should also consider the far more important fact that silent genes can mutate freely along an almost infinitely large number of pathways, over long periods of time and without deleterious effect on the population of host organisms.

After numerous mutations have accumulated in the silent gene in question, as and when the gene begins to code, it has no better or worse chance of being beneficial than any other macro-mutation. But that is not the point. The point is that non-coding genes offer an almost infinitely richer potential range of mutational pathways towards new variant genes, while the potential pathways of coding genes are highly restricted.

When we look into the heart of the genome, and we see how radically different are the opportunities for mutation provided by the silent genes compared with the coding genes, we begin to perceive how genomes which contain large numbers of silent genes are far more likely to evolve into varied, complex organisms.

Simultaneously, we start to perceive that natural selection of individuals is not playing the primary role in driving evolution, but a secondary role in "trimming" variation (which has already been generated) to suit local environments. According to this view, natural selection is primarily adaptive, and occurs downstream of variation — only after variation has been created by the silent genes.

The current empirical evidence

In assessing which of the two mutational sources — that of the coding genes or the silent genes — compares with the empirical data, the current evidence in 2002, when my paper was submitted to the *Journal of Theoretical Biology,* showed that there were no clear examples of coding genes mutating to form other coding genes, and four known examples of silent or non-coding genes becoming new coding genes.

Given this background, the response of the *Journal of Theoretical Biology* is hardly unexpected. It is often said of political parties who are in power for more than a decade, say, that there is a tendency to be unable to see the world from any but their own position. At the time of publication of this book, the theory of natural selection has enjoyed a 150-year predominance in biology. Unlike physics, where within living memory various theories have happily contended with one another (wave theory versus particle theory, big bang versus steady state, universe versus multiverse, etc.) generations of biologists have lived their entire professional lives under the aegis of an apparently unassailable theory of natural selection.

In a soccer match, a four-nil score would be a crushing victory for silent gene theory over Darwinism. But unfortunately this is not soccer, it is academic biology. The believers in Darwinism are not only playing on the opposing side — and virtually every player within academic biology is a Darwinist — but (because of the peer review system) the referees are Darwinists too. Furthermore, the biological journals are read almost entirely by professional biologists who are themselves convinced Darwinists. Using the soccer analogy, when one submits a new paper which questions the fundamental tenets of the theory of natural selection to a peer-reviewed biological journal, one is playing not only against the opposing team, but the referees and the entire stadium of professional biologists as well. It is easy to see how, under such a system, a four-nil defeat for the home side can be summarily dismissed and forgotten.

Since 2002 the number of examples of silent genes beginning to code has multiplied and continues to increase steadily. In addition a new frontier science is rapidly developing which is investigating the means by which silent genes control coding genes. By contrast, there is no direct evidence of one coding gene mutating into another without passing through a "silent" stage. At the time of writing, the evidence that evolution proceeds by mutations in coding genes remains, in Kimura's description, "mythic". By contrast, the evidence that significant mutation occurs in the silent or non-coding genes, and that silent genes may later begin either to code or to control coding genes, is now overwhelming.

The shape of evolution

In attempting to make progress in 2002, there were other problems to face. The referee of my paper in the *Journal of Theoretical Biology* had made the further assertion that silent gene theory is based on group selection. To understand his reasoning, it is necessary to divert briefly into the nature of this biological heresy.

Classical Darwinism specifies that the key to evolution is natural selection acting at the level of the individual organism. At various times in the past, other researchers, for example the ethologist Konrad Lorenz, have argued that there are some characteristics of living creatures, such as "altruistic" behaviour, which cannot be explained in terms of individual selection. According to this view, certain forms of altruistic behaviour, where one individual benefits another, often at direct risk to itself, cannot arise through individual selection. In explaining why altruistic behaviour occurs in evolutionary systems, Lorenz proposed the somewhat naïve hypothesis that such behaviour evolved because it was "good for the species".

At the time Darwinists such as John Maynard Smith pointed out that this was an implausible argument. In order to work, it would require a selection at species level. However, the life and death of species is a very slow and weak form of selection compared to the selection of individuals, which is far more powerful and rapid. Where individual selection (in favour of selfish characteristics) opposed species selection (in favour of altruistic characteristics), individual selection would always win out.

A great battle was fought between classical Darwinists and group selectionists on this issue, with Maynard Smith leading the charge. Most people (including this writer) would agree that the classical Darwinists won that particular battle. Later, more sophisticated theories of group selection, involving smaller "breeding groups", were put forward, and were also, for the most part, treated with hostility by Darwinists, though the argument that group selection plays a role in what is called "macro-evolution" persists, not least among ecologists and students of the fossil record.

Against this background of controversy, perhaps the most significant advance in explaining how altruistic behaviour could arise was made by W. D. Hamilton, who developed the theory of kin selection. Hamilton proposed that altruistic behaviour could be understood in terms of the genetic relatedness of individuals. For example, in any sexually reproducing species, if one's children inherited half of one's genes, it made evolutionary sense to protect those offspring, at least to the degree that the protective behaviour did not significantly reduce one's own chances of survival. Similarly, to preserve one's genetic inheritance it made sense to protect one's grandchildren (who share a quarter of one's genes) etc. Those individuals which happened, by lucky accident, to protect their children or near relations would be selected over those that did not, because larger ratios of their genes would enter future generations. Defining this relationship, "Hamilton's law" was an equation setting out the advantages of altruistic behaviour against degrees of genetic relatedness.

The beauty of Hamilton's model was that it could both "explain" altruistic behaviour within evolutionary systems, and do so within the rigorous framework of a precise system of genetic inheritance. For this contribution, amongst others, Hamilton was accorded the status of a major visionary within evolutionary theory.

Hamilton inaugurated a conceptual innovation in evolutionary theory by arguing that significant elements of the evolutionary process could be better understood by considering evolution from the point of view of the individual gene. In turn his work on gene-centred evolution gave rise to the popular literature of the "selfish gene", so fervently and lucidly argued by disciples such as Richard Dawkins. With this in mind, let us now turn towards the claims of the theory of natural selection as the most important driver of evolution.

The current status of Darwinism

Speaking for myself, I have never doubted that Darwin's theory of natural selection is a magnificent theory, rich in prediction, sweeping in its scope, inspiring in its narrative beauty. Nor have I doubted that it has been cogently developed by effective and sometimes brilliant advocates. In the context of this immensely powerful theory, why do I believe that there is a fundamental flaw at the heart of the Darwinian theory of evolution by natural selection, and that the very notion that natural selection is the main driver of evolution is mistaken?

Before proceeding once more to particulars, let us remind ourselves, if only for rhetorical reasons, that the Newtonian theories of motion provided, if anything, an even more impressive theoretical structure than Darwin's. Through their powerful defining equations, Newton's equations were astonishingly precise in their predictions. Yet in due course this virtuosity did not prevent Newton's conceptual universe from being superseded by Einstein's theory of relativity, for the simple reason that the latter was even better at describing key features of the universe.

In discussing Darwinism, we should always bear one very important principle in mind. *The fact that a theory works, and appears to work beautifully to its advocates, is no defence against the possibility that it might one day be replaced or subsumed by a more effective theory.*

For this reason, we should not be swayed by the argument that because Darwin's theory that natural selection of individuals drives evolution has predominated for 150 years at the time of publication of this work, it follows that the theory represents some form of absolute truth. The predominance of Newton's theories lasted longer still; his *Principia* was completed in 1687 and Einstein's first paper on relativity — the special theory of relativity — was published some 218 years later in 1905. Just as the predictive accuracy of Newton's theories was no final defence, so venerability did not prevent the Newtonian system from being superseded in due course by a more comprehensive and effective theory.

What is Evolution?

One of the key arguments in the paper which follows is that evolution is more than the adaptation of inherited characters. It is the grand architecture of the origins of life. To attempt to explain that grand architecture we need to do more than to point to an important evolutionary process, such as natural selection of individual organisms, and suggest that it is the central driving force of evolution.

To give an example, the equally complex process of organic decay, involving as it does the constant redistribution of elements through the ecological system, and incorporating complex biological pathways such as the Krebs Cycle, has a huge effect on the living environment. It provides a source of energy for the vast variety of organisms which feed off decay, and it strongly affects the evolution of the geographical landscape itself. Yet no one seriously claims that it is the processes of organic decay which drive evolution.

Similarly, there can be no doubt that natural selection of individuals does occur, or that it has a major effect on subsequent generations — that is to say, it is a powerful component of evolutionary processes. But is it capable of explaining the grand architecture of evolution?

It is the submission of this essay that the theory that natural selection of individuals drives the larger evolutionary process fails at a very important fence. A key element of the grand structure of evolution is the story of how simple organisms evolve into increasingly varied and complex organisms. But what generates the variation necessary to supply natural selection with its fuel? Darwin himself was acutely aware of the problem of variation — more aware than perhaps many of his modern disciples. In *The Origin of Species* he succinctly summed up the situation by writing, "... unless profitable variations do occur, natural selection can do nothing."[1]

In arguing my case, I have occasionally suggested that if natural selection of individuals were the major driving force in evolution, it would act to so decrease variation that over time the great panoply of complex life would be reduced to simple, generalised organisms. This may appear to be a largely rhetorical position. But if we look more closely at the problem, there is, I would submit, a remarkable demonstration of precisely what happens when natural selection holds sway without the significant mediation of silent genes.

According to common classification, the biological world is divided into two great kingdoms, the prokaryotes and eukaryotes. The prokaryotes are single-celled organisms without a proper nuclear membrane. They comprise primarily the bacteria, which appear in the fossil record some three billion years ago, and are amongst the oldest forms of life.

Eukaryotes, the second kingdom, have a remarkably similar genetic coding system, which indicates that the two lines were once derived from the same root. But they have a more complex cellular structure. Amongst the differences is the one that we have noted earlier — the eukaryote genome contains a much larger number of silent genes.

As it happens, all complex, multicellular life derives from the eukaryotes. Everything we can see with the naked eye, from plants to insects to reptiles and mammals, are eukaryotes. Prokaryotes, because they are tiny, single-celled organisms, are invisible to the eye.

If silent gene theory is soundly based, and silent genes are the primary cause of variation, then this is exactly the pattern we would expect. Where silent genes are present in large numbers in the eukaryote genome, the theory predicts a flowering of variation, and through variation, complexity. The great burgeoning of complex evolutionary life started approximately 600 million years ago in the Cambrian period, and it evolved wholly and entirely from the eukaryote genome.

By contrast, consider how anomalous is the position in terms of the Darwinian theory that natural selection drives evolution. According to Darwinists, natural selection of individuals is the central driving force of evolution. If so, why have more than three billion years of natural selection, acting on vast populations of prokaryotes, in a huge variety of physical environments, produced nothing more varied or complex than the single-celled organism? According to classical Darwinist principles, the existence of vast numbers of organisms with a high rate of turnover over aeons of time provides exactly the optimum conditions required for significant evolution. If this is the case, what is responsible for generating this comparative stasis? And concomitantly, what strange oddity of natural selection allows eukaryotes to generate such unprecedented variation and complexity?

Wherever we find silent genes in significant numbers, in other words, we find a great flowering of variation and complexity, and an almost inexhaustible creation of new forms. Where silent genes are present in relatively small numbers, as in the prokaryotes, even though we find all the other key conditions for evolution by natural selection, the result is a much lower rate of variation, and relative evolutionary stasis.

To my eye at least, silent gene theory appears far more powerful and precise in predicting and describing the grand architecture of evolution than the theory of natural selection. But let us now move on to a further aspect of silent gene theory. If silent genes, and not the Darwinian selection of individuals, drive evolution, what are the actual mechanisms of selection which create silent genes?

The pitfalls and byways of group selection

Commenting upon my silent gene theory on behalf of the *Journal of Theoretical Biology*, the referee stated in his second rejection that my formulation of silent gene theory depended on group selection. The referee argued that in proposing that the function of silent genes was to provide variation, I was suggesting that the group (or population) which contained individuals with silent genes would be able to survive environmental changes over populations which contained lesser quantities of silent genes.

This would be a useful criticism if it were an accurate description of the form in which I had cast a silent gene theory of evolution. But that was not what I had set out. And in an atmosphere of continuing heated debate about the efficacy of group selection, it is important to recognise the crucial difference between what I am proposing and the arguments of early or naïve group selectionists.

When formulating a silent gene theory of evolution, I was strongly aware that the theory would be vulnerable if silent genes were portrayed as solely or primarily the product of group selection. So I looked for a more powerful selection than group selection in favour of silent genes. Almost immediately I began to search, I found the solution in the selection of genes themselves.

As we shall see below, new genes are constantly recreated in the genome by various processes of random replication. If such a new gene is a coding gene, the strong probability is that it will be deleterious to the organism concerned, and therefore individuals which inherit that gene will be selected against. Thus there is a selective pressure against new coding genes. If, on the other hand, the new gene is a non-coding gene, it will not specify any physical character, and therefore it will not be subject to natural selection acting on the organism. Unlike the majority of new coding genes, it will survive in the genome, and be replicated in reproduction. By these processes of indigenous selection within the genome, a population of silent genes will slowly expand in number relative to coding genes.

If true, it follows that group selection is not responsible for the presence of large numbers of silent genes. Instead, this accidental, inbuilt selection at the level of the genes, in favour of silent genes over coding genes, is the likeliest reason for the presence of silent genes in such large numbers over time.

However, *once the silent genes exist,* the group or population of individuals may benefit from the presence of the silent genes. A population which contains silent genes is a population which is capable of generating greater variation, and greater diversity, and is therefore more capable of surviving even major environmental changes.

Summarising this argument, silent gene theory differs from naïve group selection because its main component — the selection of silent genes over coding genes — does not derive from the action of group selection but from indigenous selection processes amongst the genes themselves. Thus silent gene theory is effectively immunised from the assertion that it depends upon group selection. In asserting that silent gene theory is based on group selection, the referee was simply mistaken.

Setting these components together by stages, we begin to observe the broader outline or shape of a silent gene theory of evolution. At a certain stage in the evolutionary saga, through accidental processes, it is likely that the first inert or silent genes appeared in the early eukaryote or proto-eukaryote cells alongside the coding genes. During great lengths of time silent genes increased as a proportion of the gene population over the coding genes, and in due course even began to outnumber coding genes in the evolving eukaryote genome.

As we have outlined above, one of the accidental consequences of increasing numbers of silent genes was a much more fertile system of mutation. Silent genes can mutate by random means along infinitely more diverse pathways than coding genes, and without deleterious effects on the host organism. At a later stage, again by largely random processes, these new mutations could be "switched on" and begin to code. Variation is thus produced *sui generis*, against the variation-depleting effects of natural selection acting on the physical phenotype. *After* variation within populations has been generated by silent genes, natural selection adapts that constantly replenished range of variation to the local environment. And so the great process of complex evolution proceeds.

One of the characteristics of a successful evolutionary theory is that it should predict important features in the evolutionary landscape. Silent gene theory appears not only able to explain why silent genes are associated with a remarkable acceleration in evolution, but poses interesting further questions on precisely how that change might have taken place. For example, did significant numbers of silent genes appear before the more complex features of the eukaryote genome, perhaps giving rise to them, or after those features were already present or well established? If our theory is that silent genes are causative of variation, and thus drive complex evolution, it would further strengthen silent gene theory if future research showed that silent genes appeared in significant numbers *before* other key features of the eukaryote genome evolved. This is only one example in which silent gene theory appears far more fertile in suggesting new areas of research than neo-Darwinism.

The landscape of silent gene theory

We know from the enormous literature on the subject what the Darwinian landscape looks like. In comparison, what does the silent gene theory landscape resemble, and in particular, what are the rearrangements of evolutionary forces which it implies?

Perhaps one of the strangest aspects of silent gene theory is that natural selection of individuals, acting on its own, is a kind of poison to complex life. Organisms may be threatened by fluctuations in temperature, by local predators, or by new strains of parasites and diseases, amongst other dangers. But what if the most persistent or universal threat to the survival of complex life arises from the variation-depleting effects of natural selection acting at the level of the individual organism?

Paradoxically, it seems likely that the eukaryotes only developed into varied, complex forms because they were able to resist and effectively overcome the variation-depleting effects of natural selection acting at the level of the individual organism. In due course, I predict, we will come to see that almost the entire genetic structure of complex organisms has evolved to resist the variation-depleting effects of natural selection acting at the level of the individual organism or — alternatively expressed — as a means of harbouring resources of variation. Accordingly, we shall be able to look afresh at structures and processes such as pleiotropy (one gene influencing several characters), polygenes (numerous genes influencing one character), gene dominance and recessiveness, amongst others. Whatever their other functions, each of these features acts to *complicate* the connection between the genes and the physical characteristics for which they code, often protecting those genes from the direct, variation-depleting effects of natural selection acting on the individual phenotype.

There are other areas in which silent gene theory applies — for example in the mapping of the human and other genomes. If the theory is correct, then researchers will be looking in the wrong place for significant macro-mutation if they follow neo-Darwinian theory and search for signs of such mutation only or predominantly amongst the coding genes.

By contrast, silent gene theory indicates that the overwhelming source of mutation is likely to be the silent or junk genes, and that is where the search should be concentrated. In this sense, dissemination of silent gene theory would save unnecessary research, and direct other research towards potentially more fertile areas.

Science, innovation, and popular debate

The current situation is that no peer-reviewed scientific journal will publish the silent gene theory. After the *Journal of Theoretical Biology* had refused to publish my paper, I dutifully tried other science journals, but without success. This is a pity, but given the nature of peer-review and the way that modern professional science conducts its business, it is hardly surprising.

Curiously enough, given my own experience, I do not object in principle to the practice of peer review. It seems to me largely constructive in maintaining the quality and consistency of science. All great social enterprises contain compromises, and peer review has been characteristic of much successful and developing scientific research. Against this, it seems to me that it should be more widely recognised how obstructive peer review can become to the airing and dissemination of genuinely radical theories which actively and constructively question certain fundamental tenets of contemporary scientific beliefs.

This predicament is not entirely without its advantages. One of my hopes about any evolutionary discussion of silent gene theory is that in due course the debate should not only be the province of merely "self-interested" parties — that is to say of professional biologists who have a lifelong personal and ideological investment in Darwinism. In the broader scientific world I hope that the jury will include — amongst others — physicists, who have their own distinct evolutionary theories of the universe, and geographers, who also are familiar with evolutionary theories (for instance, in the form of continental drift and tectonic plates). These and other scientists would be able to supply a disinterested judgement on the relative efficacy of both the rival grand theories of biological evolution under consideration. And beyond these disciplines lies the vast, wider readership of those who are simply intrigued and fascinated by the question of the evolutionary origins of life.

If, in the intermediate term, silent gene theory is not to be debated by scientists through their own professional journals, then it is fitting that it should become part of that wider and more informal exchange of views which characterises our relatively open and liberal society.

Evolution, as I have argued in the opening paragraph, is the central component of our common human intellectual heritage and, like other important areas of cultural life, has benefited from non-professional involvement. Charles Darwin, after all, was not a professional scientist but an amateur naturalist — just as, in the related field of physics, Einstein was not a professional physicist but began his career as a patent clerk.

I believe Professor John Maynard Smith's description of professional science as a community bound by beliefs held in common is accurate and true. It could be argued that this cohesive structure is science's central strength. But the very structure of this community also contains its own subtle limitations. Because of its inherent configuration, professional science is superbly proficient at advancing knowledge within its own agreed parameters. But precisely because of this structure, it often appears less capable of questioning those same parameters which effectively define its activity.

The peer review system is, if nothing else, emblematic of the way that normal procedural science treats genuinely radical ideas. Because of this inbuilt conservatism, it remains to be seen by what means, and how soon, silent gene theory takes root in the scientific community. I believe that it will take root in due course — not least because it offers younger researchers more fertile and potent new lines of research. It is a theory which, in Popperian terms, is well worth testing to destruction.

Meanwhile, although the paper which follows this introduction incorporates a number of specialist terms, it is aimed also at the non-specialist reader who is interested in evolutionary theory as an explanation of the origins of life. I trust that this same interested non-specialist reader may also be able to consider for himself or herself some of the more detailed arguments in favour of a silent gene theory of evolution.

Finally, in writing the paper which follows, I am fully aware that a great deal of what I set down is speculative, and that since I am a writer not a scientist, I do not have the advantage of belonging to a community of like-minded individuals who can criticise or refine my text before it proceeds to publication. Instead, in the interests of setting out the theory in as much detail as I am capable of, my attitude has been to press ahead and take risks. In this respect I cheerfully follow Freeman Dyson's laconic dictum that "It is better to be wrong than to be vague." I recognise that some, or much, of what I have written may require modification and improvement in the fullness of time. Meanwhile, my hope rests in the general shape and direction of the ambitious theory that I have attempted to address in the more detailed essay which follows.

NOTE

If there are three words that a non-specialist reader should perhaps be acquainted with if he or she is to understand some of the more detailed arguments, they are “genome”, “genotype and “phenotype”.

The genome is the total genetic information that an organism inherits from its parents. The genotype is the genetic constitution of an organism, as opposed to its physical appearance (the phenotype).

Darwinian natural selection acts on the phenotype — the sum of the physical characteristics of the organism, such as blue eyes — in such a way that, over the generations, the genotype may gradually be modified.

A SILENT GENE THEORY OF EVOLUTION

ABSTRACT

It is the thesis of this paper that natural selection, acting at the level of the individual organism, is not the main cause of evolution, largely because it cannot explain a key feature of evolutionary processes — the general increase in variation within and between species over time. Instead, it is submitted that in any species natural selection reduces variation in favour of an optimum type for any given environment.

A better explanation of the source of variation is likely to be the structure of the genetic system itself, and in particular the fact that a significant number of the genes — those which do not code for physical characteristics — exist outside the variation-reducing effects of natural selection acting on the phenotype. These "silent" genes are able to evolve more freely, without being "seen" directly by natural selection. Equally importantly, because they do not code for physical characteristics, the mutations they acquire over time are not deleterious to the host organisms. Silent genes can thus gradually accumulate highly exotic multiple mutations. When eventually these mutant genes are switched on and begin to code, they are sifted by natural selection in the normal manner.

Greater numbers of silent genes than coding genes, freer pathways of mutation, and non-deleterious effects on the host organism while in the silent stage all combine to create a far more powerful system for generating effective mutational variation in the silent genes than in the coding genes.

Within the genome itself, a new selective process is proposed which selects in favour of silent genes. While natural selection acting at the level of the phenotype acts against deleterious coding genes, new silent genes are created by random replication. As the ratio of silent genes gradually increases over time, so does the rate of evolutionary development.

Empirically, across a whole range of predictions, silent gene theory corresponds much more closely with the newly emerging research data than the theory of natural selection. Until recently it was thought that the number of coding genes would be correlated with species complexity. But sequencing of the full genomes of a wide range of species shows this relation to be weak, and replete with major anomalies. For example, primitive nematodes have more coding genes than insects, and rice has more coding genes than humans.

In direct contrast, and as predicted by silent gene theory, there is a far more powerful correlation between the ratio of silent genes in the genome and species complexity. Silent genes are present to some degree in prokaryotes (5%-24% of the genome) but their ratio in the genome increases markedly in eukaryotes in proportion to the complexity of the organism. Nearly all multicellular species have more silent genes than coding genes, the ratio of silent genes rising in a smooth curve with complexity of species to a high of 98.5% in the human genome. There appear to be no major anomalies in this correlation.

The general rule proposed in this paper is that silent genes generate the great majority of variation across the whole panoply of evolutionary species. By contrast, natural selection, because it acts to reduce variation, operates mainly as a negative or entropic force in evolution. As a consequence, many of the detailed structures and processes of the genome are best understood in terms of their capacity to resist and overcome the variation-depleting effects of natural selection.

Since the main generative process of complex evolution is the variation generated by the silent genes, it is this, and not natural selection, which is the main driver of evolution.

Background

Darwin's theory of natural selection of individuals is perhaps the greatest and most durable intellectual edifice of our time. It is an inspiring and beautifully argued thesis, and its progenitor is rightfully regarded as the founder of evolutionary science. In explanatory terms, the theory of natural selection has not only decisively defeated its historical rivals, such as Lamarckism, but has established biology as a major empirical discipline. Under its broad auspices genetics has expanded into fields of research which promise to bring massive human benefit.

All knowledge is tentative, however, and such remarkable success does not mean that natural selection as a theoretical system is infallible, or that certain features cannot be improved. This paper is an attempt to explore what seems to be a consistent fault line in current evolutionary theory — the continuing attempt to define the true or main source of variation — and to supply a tentative solution.

In what follows, it is perhaps useful to begin by distinguishing between the theory of natural selection and the physical processes of evolution. Darwin proposed that evolution is driven by the natural selection of individuals, and in particular that those individuals which are better adapted to their environment propagate themselves at a differentially higher rate.

Natural selection is a very powerful and pervasive process, but despite its significant effect on the evolutionary landscape, this is not the same thing as driving evolution. One of the central characteristics which a grand evolutionary theory must explain is how simple, generalised organisms have developed into more varied, complex organisms. Such an evolutionary theory must postulate, in other words, the cause of increasing variation both within species and in generating speciation. If we examine closely the process of natural selection, however, we see that it acts against extremes of variation in favour of an optimum adaptation for any given environment. In other words, *the general direction of natural selection is counter-variation.*

Proponents of the theory that natural selection drives evolution argue that the environment is heterogeneous, and that active adaptation to a heterogeneous environment generates increasing variation by virtue of adapting to that heterogeneity. In key respects, however, this argument appears to be an inversion of cause and effect. The heterogeneity of the environment depends in overwhelming part on the variety of organisms within it, and the thesis that environmental heterogeneity causes variation seems perilously close to the tautology that variation is itself the cause of variation. Instead, it is incumbent upon evolutionary theory to explain how that heterogeneity developed, before ascribing it as the primary cause of variation. Meanwhile, it appears more logical and more likely that variation is the cause of environmental heterogeneity, not its effect.

There seems, however, a potentially more lethal argument against the view that heterogeneity of the environment is the "cause" of variation. Let us assume, for the purposes of argument, that the environment is infinitely heterogeneous. No matter how heterogeneous the environment, in every niche in the potentially infinite number of niches in that environment the same process is taking place. Variation is always being reduced by natural selection in favour of optimum adaptation.

If we were to describe, intuitively and *a fortiori*, a system in which natural selection were the predominant mechanism of evolution, it would be one in which variation in any species was always in the process of being reduced in favour of an optimum type for any given environment. Such continuous reduction in variation in favour of optimum environmental adaptation should eventually reach a point at which, for any given species, if the environment changed significantly, there would not be sufficient resources of variation within that species to adapt, at which point the species in question should die out. Far from being "the origin of species", as Darwin proposed, natural selection of individuals, acting on its own, seems more likely to cause the extinction of species.

There is, it is generally agreed, a natural and high rate of species extinction in the evolutionary system. The average life-span of a species is only about 4 million years. Because of this, there are a vastly greater number of dead species than living species. Given this background, it could be argued that we need a powerful and comprehensive theory to explain the natural, gradual and orderly processes of extinction. It may be suggested that Darwinian natural selection provides just such an explanation. If this is so, then natural selection, properly defined, would still serve an important theoretical purpose within general evolutionary theory, though not one which Darwin had necessarily intended or foreseen.

Amongst the biological community, more recent developments of Darwin's theory are held to have increased its explanatory and predictive powers. For example, the concept of gene-centred evolution, originally developed by W. D. Hamilton[2], and persuasively applied to a variety of phenomena by brilliant practitioners such as J. Maynard Smith, E. O. Wilson, R. Trivers, R. Lewontin, C. Williams and others, appears to offer certain improvements upon Darwin's theory of the natural selection of individuals. As a means of explaining differential rates of genetic inheritance, including precise and testable accounts of such aspects as altruistic behaviour, these gains are powerful and undeniable. However, the chief difficulty in the concept of the "selfish gene" as a revised Darwinian evolutionary theory is that it merely transplants natural selection from the individual organism to the individual gene. It does not cite "silence" as a key factor. In this sense, gene-centred evolution suffers from the same significant criticism as that levelled above against the natural selection of individual organisms — namely, that the genes which code, and which are therefore associated with phenotypic characters, would be acted upon by natural selection to reduce overall variation in the gene-pool. If this process were predominant, the genetic variation within a species should be gradually but inexorably reduced until the environment undergoes a significant change; at such a time the species, depleted of variation, should be unable to adapt to significant change and is likely to die out.

Taking into account these preliminary doubts, let us consider further attempts to meet the problem of precisely how variation may be generated.

Further advances in evolutionary theory

Biologists have long recognised that natural selection reduces variation, and over the years an attempt to resolve the problem has been proposed in the form of the "modern synthesis" (also sometimes called neo-Darwinism). The key element of the modern synthesis — that variation is caused by mutation amongst those genes which code for physical characteristics — seems to combine a significant conceptual advance with an equally significant empirical weakness. The conceptual advance is that the theory recognises that the source of variation is not natural selection but derives from processes within the genes themselves ("mutation"). This insight will be addressed more comprehensively as the paper proceeds.

The empirical weakness at the heart of the modern synthesis can be outlined in relation to the eukaryote genome, which in turn is the base of more "complex" evolution, including all multicellular organisms. The modern synthesis postulates that the chief source of mutation is the coding genes, the genes which code for proteins and thus for physical characters. According to the modern synthesis, non-coding genes are extraneous to this process and consequently were called "junk" genes. This weakness will also be considered further.

Meanwhile, there are three reasons for believing that mutation within the coding genes represents an inadequate source of variation. The first is that the genes which code for characters represent a minority of the DNA in the eukaryote genome — in *Homo sapiens* this figure is only 1.5%. Secondly, the theory of natural selection dictates that mutational advances must be achieved along pathways which are themselves functioning genes (see below). This greatly limits those mutational pathways. Thirdly, these same coding genes which supposedly generate variation by occasional mutation are themselves subject to the powerful counter-variation effects of natural selection acting on the phenotype.

For these three reasons, it seems more likely that the relatively small amounts of mutation generated by genes which code for physical characters will be swallowed up by the powerful variation-reducing effects of natural selection (see below).

Before, however, we proceed to explore more likely or plausible sources of variation in natural systems, there is one further theory of evolution which should be considered.

The neutral theory of evolution

Starting in the early 1950s, the population geneticist Motoo Kimura produce a celebrated series of papers on mutational change[3]. Kimura demonstrated that the frequency of mutation — in which mutations are defined as single base changes — was much greater than expected. A very high proportion of such mutations were shown to have little or no effect on the phenotype. Kimura is widely admired for his empirical work in identifying the high numbers of base changes/mutations amongst the genes, and for his strong development of Haldane's and Sewall Wright's notion of genetic drift as an important principle in evolution.

In 1968, departing a little further from this empirical platform, Kimura also began to set out what he called the neutral theory of evolution[4]. This theory proposed that, since the great majority of mutations (considered as base changes) have little or no effect on the phenotype, the main course of evolutionary advance is along a "neutral" axis. From our perspective, the neutral theory is something of a milestone because it represents a significant attempt by a distinguished researcher and his colleagues to address the problem that natural selection acting on the phenotype may not be the main driving force in evolution.

At first Kimura's reputation as a population geneticist, and the novelty of his proposal, caused a favourable, or at least tolerant, reception to his theory. Slowly, however, resistance began to build to the neutral theory of evolution. Amongst the biological community, the processes of natural selection and physical adaptation of the phenotype appeared to be so clear and powerful that the notion of neutral evolution as the main driving force of evolution seemed, if not flawed, then almost tangential to their own concerns. In addition, it did not seem to those versed in natural selection that Kimura's theory corresponded with a variety of empirical data.

One of the key difficulties of dealing with Kimura's theory is that his interest in neutral base substitutions which take place without affecting the phenotype has virtually no point of contact with the evolving neo-Darwinian theoretical structure, which is based on continuous phenotypic modification by natural selection. This may be illustrated by the following thought experiment. Suppose that we have the good fortune to meet two distinguished population geneticists, Kimura 1 and Kimura 2. Kimura 1 is the one we know. Kimura 2 is another distinguished molecular biologist who looks exactly the same, and is also the author of the neutral theory of evolution. The only difference between them is that in the case of Kimura 2 so many neutral substitutions have taken place that there is not a single molecule in common with Kimura 1. To Kimura 1 the two beings are entirely different — 100% different, in fact. To the rest of us they are the same.

Like all good scientific theories, Kimura's neutral theory of evolution was capable of being tested. The predictions which could be drawn from such a well-defined genetic "drift" theory, not least the dispersion patterns which could be extrapolated from it, were shown not to be borne out.

As evidence has mounted against his theory, Kimura has not publicly disowned the neutral theory of evolution, but since 1986 he has not publicly defended it or advocated it either[5], and it is widely believed to have subsided as a serious alternative to the neo-Darwinian theory of continuous phenotypic modification. Meanwhile, Kimura retains a highly deserved place in the biological pantheon for his outstanding work in population genetics and for demonstrating rates of point mutation in diverse populations.

Silent gene theory

A major disadvantage of Kimura's neutral theory of evolution is that it fails to solve the central problem of the cause of variation. In proposing that evolution proceeds along a neutral axis, the neutral theory does not address a key question; if natural selection acts to reduce phenotypic variation, what precisely is the main source of such variation? In considering this critical question, let us instead attempt to consider the problem head-on, within the classical framework that evolution proceeds by natural selection acting on the phenotype.

Part of the inspiration of silent gene theory is the adventurousness and audacity of physicists in proposing the presence of the gravitational influence of "dark matter" as the reason why parts of the universe expand at a slower than predicted rate. Taking account of the above, silent gene theory began as a series of speculations in an attempt to locate a powerful new source of variation within the genome, preferably one which could resist and even overcome the variation-reducing effects of natural selection. Might it be that a hypothetical special class of genes — so-called "silent" genes — are capable of mutating while being, at the same time, partially or wholly immune from the counter-variation effects of natural selection acting on the phenotype? And could these genes contribute to phenotypic variation in due course by emerging from silence

— through largely random processes — and coding for physical characteristics?

To conform to these theoretical requirements, silent genes would need to entail one further important characteristic. As a source of significant mutation, they should be present in very large quantities in the genome. When expressed theoretically, these seem highly exotic requirements.

It is these odd speculative requirements, however, which increasingly directed our attention towards the so-called "junk" DNA. In two important areas, junk DNA seems to conform closely with our requirements. Because junk DNA does not code for proteins, this means that natural selection on phenotypic characters does not affect it, at least while it is silent. Secondly, there is increasing evidence that junk genes — those areas of the genetic code which are inert — may be switched on and begin to code for physical characteristics. We will discuss this in more detail below. Thirdly, because junk DNA constitutes typically a major proportion of the genome in complex eukaryotes, rising to 98.5% of the genome in humans, it is certainly present in sufficient quantities to provide significant mutations.

But once the broad parameters of silent gene theory have been set down, there is a further characteristic of silent or junk genes which suggests that the theory may provide an answer to the perennial problem of how variation may be generated within the genome despite the prevalent counter-variation effect of natural selection acting on the phenotype. If such silent DNA exists outside the direct purview of natural selection acting on the phenotypic characters, it is capable of evolving along pathways which can include non-functioning intermediate stages *without deleterious effects on the host organism.* This in turn opens up rich new pathways towards new genes, a subject which will be described in more detail below.

The origins of silence

How do silent genes form, and how do they perpetuate themselves?

The processes of formation are largely known and identified. Random replication of genes appears to play a large part in their formation. Silent genes accumulate over time in the genome precisely because, unlike coding genes, they are not themselves subject to the winnowing processes of natural selection acting on the phenotype. If a random replication produces a new mutant coding gene, the chances are that its phenotypic effects will be deleterious, and the organisms which inherit the new coding gene will be selected against over time. However, if the new gene is silent, it has no direct effect on the phenotype. Instead the silent gene will continue to be replicated in the normal reproductive process without being affected by natural selection acting on the phenotype.

This means that, relative to coding genes, on average silent genes last longer. Such a selection process in favour of silent genes, acting over long periods of time, appears to be the most likely explanation for the increasing preponderance of silent genes in the eukaryote genome.

In neo-Darwinian terms, the self-replication of silent genes was thought to be merely an example of selfish behaviour at the level of the individual genes. This assumption appears questionable, for three reasons. The first is that "parasitic" activity on this staggering scale simply does not make sense, not least in terms of the theory of natural selection of individuals, which suggests that every molecule within an organism is sifted for its contribution to the individual organism's survival and propagation. Secondly, the effort of replicating a body of silent genes which (in complex eukaryotes) can be more than 50 times larger than the coding genes seems an extraordinary addition of unnecessary complexity to a genome *if such silent genes play no constructive part in its survival.*

A third objection follows from the first two and is to some extent derived from them. Only if the silent genes are playing some (hitherto unspecified) function does it seem likely that they would proliferate and survive in such numbers.

The reason for the huge preponderance of silent or non-coding genes appears to be that they are indeed playing some vital constructive role in the longer term survival and propagation of the genome. If this is so, then the processes of gene selection, and the longer term benefit to a breeding population are probably aligned, not conflicting.

As this essay progresses, an attempt will be made to build up a typical or likely biography of the life of a silent gene. But first let us consider the function that they may perform.

An alternative theory of evolution

To what extent are silent genes not merely "junk" DNA, but a kind of *de facto* factory for the creation of new variant genes? According to this view, the silent genes are like an evolving dictionary — of letters, words, phrases and sometimes full paragraphs or chapters — which can be used or reintegrated in due course as part of the evolving genetic machinery. Like words or phrases in a real dictionary, a large number may fall into disuse, but a significant number may also be created afresh and become incorporated over time into the operative language.

In its broad outline, this may seem at first like a form of group selection, and therefore it is worth at least outlining the selective components involved. Silent genes are retained in the genome not because of the long term "group-selective" advantage they may donate. Rather, their presence is the direct result of an intra-genomic selection between different classes of genes. Because they are not subject to the winnowing effects of natural selection on coding genes, they tend to last longer than coding genes. Given that they subsequently accumulate in large numbers, an accidental or unintended consequence is that they provide the base for the future evolution of genomes.

It would appear then that two processes — intra-genomic selection in favour of silent genes, and the longer term evolutionary advantages of the presence of silent genes in the genome — are aligned and mutually reinforcing. Against these two selections is a selection at the level of the individual organism which acts to reduce unnecessary or non-functioning genes within the genome. The result of these three balancing selections — two in favour of silent genes, one against — is a large "working population" of silent genes within the genome, sufficient to generate continuous variation and to power complex evolution.

If this is so, it provides at least a preliminary answer to a prescient question posed as early as 1970 by J. Maynard Smith[6], who investigated the theoretical notion as to whether and how a protein might pass through various mutations on its way to a new functional configuration. As Maynard Smith argued, neo-Darwinian theory suggests that, in order to survive natural selection of phenotypic characters at each stage, each of the intermediate phases must also be functional, and yield survival benefit. In pursuit of this investigation, he proposed a tentative theoretical model by which such evolution might be possible (see below).

The same principle may be applied to the genes themselves. But the necessity of confining pathways to functioning intermediate stages seems to load probability against the chances of effective and smooth evolution. By contrast, the notion of a large number of silent genes existing outside direct natural selection acting on the phenotype implies that such genes could pass through a number of non-functioning mutations before a functional configuration is reached.

In this important sense, silent gene theory is almost the opposite of Kimura's neutral theory of evolution. The final products of silent genes are not genes which are neutral, but genes of potentially high positive or negative survival value, created by processes lying outside direct natural selection acting on the phenotype. At a later stage, these same genes are precipitated into the process of natural selection of the phenotype when in due course they start to code.

Like mutations generated in the coding genes, when silent genes begin to code, the great majority are likely to be deleterious to the organism in question. But this is not the point at issue. The important point is that silent genes can create not only a far greater quantity of new mutations (especially where they outnumber coding genes in the eukaryotes) but, more importantly still, because they can mutate freely along almost infinitely more complex pathways, they can produce a far greater richness and variety of mutation than the coding genes. *Significant mutation is therefore overwhelmingly more likely to derive from the silent genes than the coding genes.*

This is the alternative picture of evolution which we deduce — of a slow but vast accumulation within the gene-pool of silent genes, distanced from direct natural selection of the phenotype, whose gradual mutation, random re-arrangement and eventual precipitation out of silence in due course creates the great majority of variation in eukaryotes. Such enhanced fertility of variation becomes in turn the driving force of eukaryote evolution.

In the sections which follow, this general outline will be adumbrated in more detail, including a consideration of the evolution and structure of the genome itself. First, however, let us consider in more detail the methods by which such new genes are likely to be formed in the silent part of the genome, listing these aspects point by point. As we shall see, the "inert" or silent part of the genome may be far from inactive when considered from other perspectives.

The pathways of gene evolution

At first sight, the interior of the genome appears to include a bewildering array of such entities as transposons and retrotransposons, satellites and minisatellites, pseudogenes and retro-pseudogenes. Many of these entities are characterised by the ability to replicate themselves by containing instructions to create copies of themselves. A transposon, for example, is a gene sequence varying from 750 base pairs to 40 kilobase pairs, which is capable of moving from a site in one genome to another site in the same or a different genome, leaving its original intact. A satellite is a tandem repeat of a DNA sequence; it can evolve rapidly and can shift position. A pseudogene is a DNA sequence which, despite resembling a functioning part of a gene elsewhere in the genome, does not code for proteins, and may have resulted originally from a replication. Pseudogenes, often having existed in a silent state for a considerable time, are likely to have accumulated a number of further mutations.

These classes of genes demonstrate that there is a great deal of apparently random replication and transposition within the genome, much of it resulting in various types of rearrangement of the genetic material. In causing or fomenting these changes, entities such as transposons and satellites appear to be amongst the most active agents of change.

From a neo-Darwinian perspective, focused on the coding genes and the competition between fit and less fit coding genes, nearly all of this self-replicating activity amongst silent genes must seem like selfishness running mad; hence the use of the word "junk" DNA to describe the general category of non-coding DNA, and the amiable but telling use of such terms as "pseudogenes" to describe entities which are perceived to have no useful function in relation to the "real" genes, the active genes which code for physical characters.

In terms of silent gene theory, by contrast, a more complex and perhaps more constructive picture emerges, one in which a licensed freedom operates in the silent part of the genome, resulting in a remarkable set of interactions whose final product is a range of new functioning genes which are unlikely to be produced by other methods (see below). In the course of these processes, transposons and satellites act as cutters and re-splicers as new arrangements of genes are made. Such activity is largely random, but — as we shall argue below — one consequence is that it greatly facilitates the formation of new genes. Although the purpose of this paper is to outline silent gene theory as a general alternative to the modern synthesis, a closer look at the remarkable activity within the silent part of the genome may serve to illustrate more sharply certain vital differences between the two theories.

As outlined earlier, Maynard Smith[6] investigated the theoretical notion that in order to reach a new configuration, proteins must pass through a series of intermediate stages which are themselves functional. Using word games as an analogy, and functioning words as representative of functioning proteins, he demonstrated the type of constrained path by which Darwinian evolution could proceed. For example, he showed that by means of single substitutions of letters, each substitution representing a mutation, it was possible to move from WORD to GENE by passing through four separate functioning sequences:

WORD ⇒ WORE ⇒ GORE ⇒ GONE ⇒ GENE

This is an analogue of evolution, in which the words represent proteins, the letters represent acids, etc. Such a demonstration clearly elucidates both the nature of the possible paths and (at the same time) the profound limitations of neo-Darwinian evolution of the coding genes. What applies to proteins also applies, in a broader sense, to the genes which code for them. In particular, since the neo-Darwinian pathway of mutations consists of single mutations within coding genes, this constrains the coding genes to sequences of approximately the same length. By contrast, one of the obvious characteristics of genes is that they have prodigiously different lengths, from 20 base pairs on a minisatellite to 40,000 base pairs on a more complex gene.

In what follows, we will use human language as a loose analogy of the genetic language, in order to elucidate certain processes in the formation of new configurations in the silent genes.

How, then, are these genes formed? In an attempt to develop this analogy, it is perhaps of interest to consider how a short word/gene such as CHUM becomes, by means of various gene mutations, movements or splicings, a much longer functioning word/gene — let us say, for example, CHRYSANTHEMUM. Suppose that a random copy of the functioning gene CHUM is made and that the copy, located at a different site in genome, is silent. Obviously, one of the most direct routes to a new functioning gene, corresponding to the effect of a transposon, would be if a "nonsense" gene or pseudogene RYSANTHEM were spliced into the centre of the copy of CHUM, such that that it creates CH(RYSANTHEM)UM.

In terms of the neo-Darwinian evolution of coding genes, the existence of a non-functioning gene such as RYSANTHEMUM as part of the genetic machinery is problematic. Such genes are recognised to exist in the form of "pseudogenes", but no significant or systematic reason has been given as to how they might play an active part in the genetic machinery.

In silent gene theory, by contrast, a non-functioning gene which is the equivalent of RYSANTHEM can exist for a prolonged period as a non-coding gene before being incorporated at a later stage into a copy of a functioning gene. (RYSANTHEM's non-functioning configuration may be the result of a copy of another gene which has undergone a number of changes over a period of time, and in which the presence of the coherent word/gene ANTHEM, for example, may give a clue as to its origin.)

If silent gene theory permits a number of non-functioning intermediate phases, there are a vastly greater number of different routes by which this configuration could have been reached other than those of a set of functional intermediaries (if indeed it could be created by means of functional intermediaries at all). For example, the evolution of CHRYSANTHEMUM from the gene CHUM might also consist of two random stages in which the first stage consists of a silent copy of a functioning gene ANTHEM transcribed into the middle of a silent copy of the functioning gene CHUM, giving the non-functioning intermediate CH(ANTHEM)UM.

At a later stage another silent gene, namely RYS, may be transcribed into CH(ANTHEM)UM, which in turn gives the functioning form CH(RYS)(ANTHEM)UM. The full sequence of changes may therefore be summarised as:

CHUM ⇒
CH(ANTHEM)UM ⇒
CH(RYS)(ANTHEM)UM

It is conjectured that these replications and changes represent processes which take place on a substantial scale in the genome. For example, the presence of a non-functioning gene or pseudogene, RYS, corresponds to the large number of pseudogenes which exist in the genome. (Ascribing a possible history to such a pseudogene, RYS originally may have begun its life as a copy of the functioning gene RYE, which in turn was subject to a single mutation which changed E to S.)

Alternatively, CHRYSANTHEMUM may have evolved by a third path, through three stages instead of two, such that a) a copy of a functioning gene AN may be transcribed into the centre of a copy of the functioning gene CHUM, creating a non-functioning gene CH(AN)UM; b) a copy of another functioning gene THEM may be transcribed into CH(AN)UM to create the non-functioning gene CH(AN)(THEM)UM; c) a pseudogene RYS (which in this case began life as a copy of the functioning gene RAT, say, but during its long term as a silent gene was subject to two mutations from A to Y and T to S), may then be transcribed into the silent gene CH(AN)(THEM)UM such that the final product becomes CH(RYS)(AN)(THEM)UM. The full sequence of changes over time may therefore be summarised as:

CHUM ⇒
CH(AN)UM ⇒
CH(AN)(THEM)UM ⇒
CH(RYS)(AN)(THEM)(UM)

This may at least suggest why non-functioning pseudogenes exist in considerable numbers in the genome. In neo-Darwinian terms, pseudogenes are the rusting copies of former genes, the mere detritus of earlier "selfish" acts of self-replication. In our example, although RYS is an apparently useless configuration because of the mutational change or changes it has undergone, its usefulness as a potential component of CHRYSANTHEMUM provides it with a vital role in generating a new functioning gene.

In pursuing this notion further, we submit that although silent genes are the result of intra-genomic selection, the "unintended" consequence of the creation of pseudogenes is that they evolve quietly through various mutations over time. A reservoir of such altered pseudogenes may in due course prove to be a key factor in the evolution of more complex genes. In supplying durable, non-coding (and therefore harmless) intermediate stages for evolution, we submit that silent genes are likely to provide a far richer nexus of pathways to a new generation of variant genes than the coding genes. Differently viewed, a genome which contains such silent genes, including pseudogenes, is likely to evolve more freely and faster than a genome which does not.

If genomes with silent genes generate a new range of variant genes faster and more effectively than genomes without silent genes, they and their descendants will adapt more quickly and readily to new environments. Of equal importance, perhaps, the greater variation generated will enhance protection against the continuous threat of fast-evolving internal parasites. This evolutionary accelerator effect may itself at least partially explain the functional existence of silent genes within the genome.

We have mentioned earlier that neo-Darwinian theorists, perplexed by the sheer scale of silent gene populations in the genome, have in the past described the proportion of the genome which is silent as "parasitic". From the perspective of silent gene theory, the opposite may indeed be the case. In eukaryotes the numerically predominant array of silent genes, existing at a remove from the variation-depleting effects of natural selection acting on the phenotype, appears the more viable creator of genetic variation. Compared to this, the minority of genes which code for physical characters seem to be a mere skin or sacrificial layer. This layer of coding genes appears far less able to evolve, because it is highly constrained to functioning intermediate stages, and because the limited mutation of which it is capable is likely to be overcome by the powerful variation-reducing effects of natural selection. On the contrary, it seems to us more probable that the coding genes are dependent or "parasitic" on the silent genes for their continuous replacement and evolution.

In summary, our reasoning suggests that silent genes are at the heart of the evolutionary process, and are not merely its detritus, while coding genes occupy an important, but relatively secondary, function in evolutionary development.

In what follows we will consider some additional aspects of the genome which suggest that the silent genes represent an indigenous factory for variation within the genome itself. There are at least two distinct processes by which new genes may be formed. The first is by mutational base changes, the second by means of rearrangements through re-splicing of sequences. Let us consider the latter process in more detail below.

The role of reverse transcriptase

Genes are protein recipes. One of the most common protein recipes in the entire human genome is the gene for the protein reverse transcriptase. Reverse transcriptase takes an RNA copy of a gene, copies it back into DNA and stitches it back into the genome. It is a return ticket for a copy of the gene.

In terms of neo-Darwinism the ability of reverse transcriptase to copy and re-splice genes back into the chromosome at random is considered a further text-book example of the chaos within the genome created by selfish behaviour in individual genes. It appears to be the most important process by which random copies of genes are made. Since these random copies are widely believed to be examples of entirely selfish replication, the activity of reverse transcriptase has sometimes been thought to be tangential to the survival of the genome, and perhaps even actively detrimental.

An alternative hypothesis should be considered: that the constant mixing and reshuffling of copies of genes and fragments of genes in the silent part of the genome may be an important part of the means by which new genes are constructed. As we have argued above, the primary elements of this construction process — the building blocks of what will become large complex genes — are used copies of functioning genes, copies of parts of functioning genes, and pseudogenes which began as random copies of functioning genes but have been subject to one or more mutations. These are constantly mixed and re-mixed by the action of reverse transcriptase into new sequences.

Geneticists like to point out that reverse transcriptase has another deleterious aspect because it is also the means by which retroviruses, such as HIV, stitch themselves into the genome. Against this background, it is interesting to consider whether reverse transcriptase does not have an important function in the long-term evolution of new genes.

It may be regarded as ironic, perhaps, that the strong likelihood is that reverse transcriptase was introduced originally into the genome carried in a retrovirus. On the other hand, recognised examples of the evolution of parasites into symbionts and then, later, into key functioning parts of an organism, are so wide-ranging and comprehensive in evolutionary history that the principle does not require further elucidation here. Against this background, it may be suggested that despite the possibility that reverse transcriptase may have originated in a selfish retrovirus, its propagation and extensive action within the genome may benefit the genome in certain ways. Meanwhile, let us attempt to keep an open mind on the function of reverse transcriptase, and consider a further and related aspect of silent gene theory.

Random duplication and multiple repetitions

If the silent part of the genome is the main engine of variation within the eukaryote genome, what other theoretical or hypothetical characteristics are consistent with this proposed function?

Using the example of the evolution of a copy of the functioning gene CHUM into the larger, complex gene CHRYSANTHEMUM, we have already considered how silence permits a multitude of pathways in the form of non-functioning intermediate stages in the creation of new functioning genes which are not possible in coding genes. Using functioning words as loose analogies for functioning genes, are there other clues in the genome as to how new functioning genes might be constructed over the long term?

Consider, for example, the following list of functioning words, selected from a randomly chosen text as the first twenty words which equal or exceed 11 letters in length. Like the long word/gene CHRYSANTHEMUM, in principle nearly all such long words/genes can be broken up into smaller entities which are either functioning words/genes, or non-functioning words/genes with one or more letters/sequences changed by mutation. The following initial breakdown of the first 20 words in our random selection perhaps helps to elucidate this principle:

INTOXICATION	(IN)(TOXIC)(AT)(ION)
CANTANKEROUS	(CAN)(TANKER)(OUS)
PERMUTATION	(PERM)(UT)(AT)(ION)
METEMPSYCHOSIS	(MET)(EM)(PSYCHOS)(IS)
EXTRASENSORY	(EXTRA)(SENSORY)
ORDINATION	(OR)(DIN)(AT)(ION)
CONTEMPORARY	(CON)(TEMPORARY)

INTERACTION	(INTER)(ACT)(ION)
CONCENTRATION	(CONCENTR)(AT)(ION)
MEASUREMENT	(ME)(A)(SURE)(MEN)(T)
EMBARRASSMENT	(EM)(BAR)(R)(ASS)(MEN)(T)
INDEPENDENTLY	(IN)(DE)(PEN)(DENT)(LY)
SEPTUAGENARIAN	(SEPTUA)(GEN)(ARIAN)
CONTRADICTION	(CONTRA)(DICTION)
INDIVIDUALITY	(IN)(DIVI)(DUALITY)
PHENOMENALLY	(P)(HEN)(OMEN)(ALLY)
ENVIRONMENT	(ENVIRON)(MEN)(T)
INTERPRETATION	(INTERPRET)(AT)(ION)
DEVELOPMENT	(DEVELOP)(MEN)(T)
INTRODUCTION	(INTRO)(DUCT)(ION)

While the use of human language as a metaphor for the genetic language is a loose analogy, certain basic rules seem to apply. Longer words/genes are not constructed randomly, but more generally out of smaller words or bits of words/genes. In a self-creating language such as the genetic system, it is likely that these assembly stages, in which large coherent functioning units can be incorporated, provide much faster pathways to new genes than the mere random assembly of sequences of base pairs.

But let us consider some further aspects of possible new gene assembly within the silent part of the genome. In addition to the features described above, are there other characteristics of the silent part of the genome which seem to correspond to our thesis of a factory for genes?

One of the most perplexing aspects of "junk DNA" is the very large number of replications of certain sequences. In the human genome, one of the commonest of all is a DNA sequence called LINE-1. This is between a thousand and six thousand bases long, and includes a complete recipe for reverse transcriptase in the middle. There may be as many as 100,000 LINE-1s in the human genome. According to some estimates, copies of LINE-1 are thought to account for a remarkable 14.6% of the total DNA in the human genome. Given that the coding genes are only in the region of 1.5% of DNA in humans, this is obviously a very high proportion of overall DNA.

Even commoner than LINE-1s are shorter sequences called Alus. Each Alu contains between 180 and 280 bases. The Alu sequence does not itself include reverse transcriptase but it contains instructions to other genes' reverse transcriptase to duplicate itself. The Alu text may be repeated up to a million times in the human genome, and represents up to 10% of the DNA in the entire genome.

Since, in our loose analogy, the capacity of a word to transmit meaning corresponds to the capacity of a gene to make functioning proteins, it is perhaps interesting that the Alu bears a close resemblance to a gene which codes for part of a protein-making machine called the ribosome. If the function of a coding gene is to make proteins, the copy or close copy of such a gene suggests an underlying capacity to create proteins once the gene is switched on, or when it forms part of a larger genetic combination which is switched on.

As investigation of the history of genes accelerates, we may begin to see patterns of order, or rather, of the optimisation of randomness, in the numerous repetitions of certain sequences. In LINE-1s we may observe some of the "spine" of future genes, and in the smaller but even more numerous Alus perhaps we shall see the local protein-making components of larger genes.

Is the presence of certain genes in such large numbers merely a case of selfish replication run mad? Or rather, if a consequence of intra-genomic selection is that the silent part of the genome is in fact a manufacturing system for new genes, might we not *expect* to see high orders of replications of certain sequences as part of the functioning machinery of gene construction?

The principles involved may perhaps be demonstrated using as an analogy the list of randomly chosen 20 long words above. When we analyse the words we find that they contain a surprisingly high number of repetitions of short functioning words such as ON or IN (corresponding to short functioning genes), or short nonsense words such as EN (corresponding to pseudogenes). Each of these is shown in brackets. In addition, there are longer sequences which are also prevalent, for example the 5-letter sequence ATION at the end of a word (underlined for emphasis):

(IN)TOXIC(AT)I(ON)
C(AN)T(AN)KEROUS
PERMUT(AT)I(ON)
MET(EM)PSYCHOS(IS)
EXTR(AS)(EN)SORY

ORD(IN)(AT)I(ON)
C(ON)T(EM)P(OR)ARY
(IN)TERACTI(ON)
C(ON)C(EN)TR(AT)I(ON)
MEASUR(EM)(EN)T
(EM)BAR(AS)SM(EN)T
(IN)DEP(EN)D(EN)TLY
SEPTUAG(EN)ARI(AN)
C(ON)TRADICTI(ON)
(IN)DIVIDUALITY
PH(EN)OM(EN)ALLY
(EN)VIR(ON)M(EN)T
(IN)TERPRET(AT)I(ON)
DEVELOPM(EN)T
(IN)TRODUCTI(ON)

The number of ON words contained in our list of 20 randomly chosen long words is 12, distributed over 10 words, which means that on average 50% of all the longer words contain ON. Similarly, for the nonsense short word EN, there are 12 examples in our random sample distributed over 9 words, which means that 45% of our long words contain EN. Other short functioning or nonsense words which are also highly represented in our sample are, in descending order: IN (7 = 35%), EM (4 = 20%), AN (3 = 15%), AT (3 = 15%), etc. These are very high proportions, not dissimilar to the high ratio of short Alu sequences to overall DNA in the silent part of the genome.

The longer sequence ATION appears at the end of 5 out of the 20 long words, which represents a 25% presence in our randomly chosen list. It may be argued that the reason ATION appears so consistently in our long English words is that it originates from the Latin nominative form *atio* at the end of the word, and that its presence merely illustrates the particular roots of the English language. But this helps, in a certain way, to make our point. There may be basic sequences which are useful in completing functioning genes just as there are fixed sequences of letters which appear useful in completing a significant number of functioning English words. It would increase the probability of that sequence being spliced into evolving new genes if that sequence replicated and spread itself through the genome in large numbers. This may give an indication of the possible reason for the very high prevalence of LINE-1 — that is to say, LINE-1 may function as a potential staple in the formation of new genes.

The neo-Darwinian belief that non-coding genes are random copies of coding genes may distort the perception of cause and effect within the genome. Perhaps, instead, close resemblances between coding and non-coding genes are also due in certain instances to a history of coding genes deriving from common sequences of non-coding genes. This too requires careful and open-minded investigation as the detailed study of the human and other genomes proceeds.

In contradistinction to the neo-Darwinian assertion that the presence of certain repeated sequences in such high ratios constitutes *a priori* proof of selfishness run mad, in future research the proposed utility of often-repeated sequences in the silent part of the genome should be examined more closely, in particular by searching for evidence of their large-scale incorporation or integration into functioning genes during the evolutionary process.

Silent genes in non-eukaryote species

Although we have deliberately confined our argument thus far to silent or non-coding genes within eukaryote species, an important empirical objection to the theory that silent genes generate variation, and so drive evolution, is that there seem to be a large number of organisms, particularly in the prokaryotes, which appear to have adequate evolutionary potential without maintaining genomes consisting largely of non-coding DNA. Prokaryotes, which include such simple unicellular organisms as bacteria, have a simpler genetic structure in which the number of silent genes is a significantly smaller proportion of overall genome size than in the eukaryotes (across the range of prokaryotes the ratio of silent genes to coding genes is thought to be between 5 - 24% of the overall genome). Prokaryotes also lack the nuclear membrane separating the genome from the rest of the cell which is characteristic of eukaryotes.

One way of treating this problem is to attempt to "immunise" silent gene theory by specifically confining it only to eukaryote evolution. In practice, our argument builds in this provision by constantly referring to the kind of "complex" variation of which only eukaryote genomes seem to be capable — such as the evolution of true multicellular organisms. This is a reasonably clear distinction. The entire vast kingdom of prokaryotes has not produced a single differentiated multicellular organism — a fact which surely requires explanation.

However, because our interest is to assess the potential universality of silent gene theory, let us consider briefly whether there may be other mechanisms of indigenous variation in prokaryotes which parallel that of silent genes in eukaryotes. For example, the capacity to produce various kinds of long-lasting spore stages appears widespread in prokaryotes. Such spore stages might be considered to correspond (in a broad or metaphoric sense) to a reservoir of silent genes in two important senses: (a) that the genes in a spore are not engaged in coding for active life processes, yet (b) they are at the same time capable of mutation along free paths not affected by natural selection of the phenotype. When they cease to be spores, and return to active life, after time periods which often correspond to numerous generations of the active form, they may bring new sources of variation to the population.

It should be emphasised that this hypothesis is not based upon a group-selective explanation. In constructing such a model, it is assumed that the disposition of active and spore stages in the population is a matter of individual selection, in which individuals respond to local conditions to optimise their survival. "Active" stages could be more applicable in locally benevolent conditions; spore stages in more difficult conditions, where sheer individual survival takes precedence. The consequence, however, is that there would exist, in that population of prokaryotes, a reservoir of "silent" genes in the spore stages, outside the usual processes of natural selection acting on the phenotype, but capable of mutating and adding to the evolutionary potential of that population.

This could lead to further predictions. For instance, it would be interesting to see whether there is more rapid evolution in prokaryotes with marked spore stages than those which do not have spore stages, or those in which spore stages are less prevalent. It might also be useful to more closely investigate the relation between spore stages and other mechanisms, including various forms of recombination, which also generate indigenous variation.

Using a parallel illustration amongst eukaryotes, in nematodes and tardigrades, for example, there is also a capacity to dry out and so survive difficult conditions. As Richard Ladle[7] has demonstrated, a fascinating pattern emerges. Those species that do not dry out are sexual. Those that can dry out are all-female.

Ladle's proffered explanation is Hamiltonian. He proposes that drying out is a means of purging parasites, and that in the other species sexual reproduction is an alternative strategy for generating that variation necessary to keep pace with parasite evolution. But others, for example

M. Ridley[8], have pointed out that parasites such as viruses should in theory have no trouble in surviving drying out as well.

There is another explanation of this pattern — derived from silent gene theory. One of the random or accidental consequences of drying out is that it generates inert or inactive genes which correspond to "silent genes"; that is to say, genes which are not actively coding, yet may nevertheless mutate. This in turn provides an alternative source of variation.

Does this mean that drying out provides a source of variation which in other nematode and tardigrade species is supplied by sexual reproduction? The answers to these and other related questions will supply useful evidence of the degree to which neo-Darwinism and silent gene theory are capable of explaining key evolutionary developments lying outside the evolution of the eukaryote genome.

One of the aspects of prokaryote evolution which weakly shadows or prefigures the effects of silent genes is the ebb and flow of subpopulations into smaller isolated environments. The dynamism of certain physical environments (for example, the formation of rock pools on sea or lake or river shores) has the effect of siphoning off sub-populations into conditions of relative stasis, followed by their reintroduction (when the rock pools rejoin the common sea or lake). Divergent forms of evolution may occur by random drift in these isolated subpopulations, and the result of their rejoining the main population (say when a high tide returns, or the lake or river floods) injects a degree of variation into the main population. By such random means variation and thus survivability might be improved in the main population.

It should be emphasised that the results of such conditions are likely to be relatively weak in evolutionary terms. A much more powerful revolution in variation occurred when silent genes appeared in significant numbers within the genomes of individual organisms. Inert "pools" of genes began to occur in the genome itself, able to evolve outside the evolutionary pressures of natural selection acting on the phenotype. It is our prediction that such pools of silent genes began to appear in certain prokaryotes. At such time a proportion of variation began to be created *sui generis*, and so the rate of evolution began to increase. This in turn led to the formation of the eukaryote genome, to increasing ratios of silent genes, and in turn to the explosive evolutionary development of the Cambrian period.

Such a digression may help to demonstrate, in a preliminary sense at least, that silent gene theory does not merely apply to silent genes within the eukaryote genome, but potentially to other species and systems in which genes happen to be "removed" or "suspended" from actively coding for physical characters, during which phase they can nevertheless mutate and so contribute to future variation. The central point which these speculations are intended to illustrate is that even outside evolutionary systems driven by large numbers of silent genes, processes may have evolved which parallel the effect of silent genes, and help to mitigate the variation-reducing effects of natural selection.

Although a range of 5-24% is a relatively small proportion of silent genes to coding genes in prokaryotes, it does permit some degree of indigenous variation. Further examination of the evolutionary record, particularly through sequencing of prokaryote genomes, will reveal to what degree prokaryotes may have evolved in proportion to the ratio of silent genes contained within their genomes. Did a stream of prokaryotes give rise to the first features of the early eukaryote genome. And if so, was the originating stream of prokaryotes characterised by an unusually high proportion of silent genes?

In other ways, too, silent genes may have played an indirect but significant role in the evolution of prokaryotes. Eukaryotes — particularly complex multicellular species — have generated a vast number of new niches capable of being exploited by parasitic and symbiont prokaryotes. Those prokaryotes which are capable of adapting to the new niches offered by rapidly evolving eukaryotes should, according to silent gene theory, exhibit greater variation than other prokaryotes. This suggests in turn that those prokaryotes capable of exploiting such niches should have larger ratios of silent genes. Such reasoning presents another potential test of silent gene theory compared with the theory of natural selection. It would be a further confirmation of silent gene theory if the prokaryote parasites and symbionts of eukaryote species were found to exhibit significantly higher ratios of silent genes than the prokaryote average.

In what follows, we shall return to a direct consideration of the eukaryote genome. This is partly because eukaryotes represent to date by far the most powerful and wide-ranging efflux of complex phenotypic variation. In addition, certain features of the eukaryote genome (other than the large amounts of silent DNA) appear to be significant in terms of their contribution to indigenous variation. These aspects will be addressed below.

The larger structures of silence

In the evolution of the eukaryote genome, it is postulated that a significant proportion of genes were originally selected intra-genomically for the property of silence, because silent genes last longer on average than coding genes. In seeking to define the physical means by which the process occurred, random gene duplication — that is to say, tandem duplication along a chromosome — is likely to have played a major role. A consequence of duplication is that one copy is free to mutate and take on a new function, the other functioning as normal. According to silent gene theory, this randomly created additional copy is therefore likely to survive longer by virtue of the fact that it is not subject to natural selection acting on the phenotype.

The result is a slow accretion of silent genes in the genome, until such time as random mutation, or other genomic processes, assigns these genes a coding or more indirect role in influencing the phenotype.

It is widely accepted that species evolve by the gradual copying, modification and recombination of existing genes, rather than by radical leaps. A gene which specifies one character or function may end up, with only limited modification, generating another character or function by means of slightly altered proteins. To give only one example, a protein called a lens crystallin helps to build the lens of an eye; yet it has virtually the same structure as alcohol dehydrogenase, the protein that digests alcohol in the liver. Indeed, the eye protein is so similar that it can degrade alcohol in a test-tube. Using this example, the broad pattern we suggest is that a gene may be randomly duplicated, and exist in its duplicated form as a silent gene for a considerable time — enough to undergo one or several further mutations and/or become spliced with other silent genes. In due course it may emerge as an active gene, often in a wholly different context, engaged in generating proteins for a new function as widely divergent from its original function as a protein for the lens of the eye may be from one that digests alcohol in the liver.

Part of the accepted evidence in favour of substantial gene duplication and re-integration in different form is believed to be contained in multigenes. Genes with considerable base sequences in common are thought to have descended from a single ancestral gene through gene duplication and modification. The common base sequences are believed to be homologous and conserved. It is also widely agreed that such multigene families are represented by genes encoding globins, immunoglobins and nuclear receptors, by homeobox- and paired box-containing genes and by genes whose products contain multifinger loops.

This evidence offers an initial insight into the strange world of the genome, and in particular into the complex of genomic and intra-genomic selective processes at work there. The incorporation of a randomly copied gene into a multigene suggests a particular form of emergence from silence, one in which a silent gene may be "co-opted" into the structure of an emerging multigene, rather than undergoing an indigenous change or mutation which causes it to precipitate itself from silence.

This process of "co-opting" a silent gene seems to square with the peculiar logic of intra-genomic selection. If it is the case that silent genes last longer on average than active genes, then it is not in the "interest" of a silent gene to precipitate itself out of silence, to start coding for proteins, or to begin affecting the phenotype in other ways, and thus to subject itself to the winnowing processes of natural selection acting on the phenotype. It may be instead that another gene or other genes co-opt or "press-gang" a silent gene into service. Indeed, we may find in due course that one of the mechanisms by which silent genes emerge from silence is through some form of co-option or "switching on" by another gene or groups of other genes.

Overall, the longer-term increase in silent components yields a genome which has silence imprinted on many of its structures and processes. One of the possible evolutionary consequences of diploidy, for example, as opposed to haploidy, is that with two functional representatives of a locus, it may not matter if one loses its original purpose. Much of the DNA in a diploid structure is not expressed, and is therefore silent. The phenomenon of dominance and recessiveness within a diploid structure may also be considered as a means of hiding or masking the effects of the recessive genes (usually more recent mutations) from the full force of natural selection acting at the level of the individual organism.

It is our thesis that in evolutionary history the advent of silent genes, and the resulting genetic structures and processes which evolved with them, thus provided a new and more powerful agent of evolution in the form of a developing genome which was increasingly capable of resisting and even overcoming the variation-depleting processes of natural selection. This new genome could "hatch" significant novel variation through its reservoirs of silent genes, and could introduce such variation to the forces of natural selection with some degree of protection or "masking", for example in the form of recessive alleles which, in the heterozygous state, are "silent", and only in the homozygous state are subject to natural selection.

The degree to which the genome incorporates processes which in turn appear able to manage and control the components of variation is also reflected in the structure of loci, within which fixity a variety of alleles can be tested against other alleles through natural selection.

Natural selection is always a vital component of evolutionary systems, but by these and other methods the genome could begin to confine the effects of natural selection to those aspects of its function and internal processes which would enhance its own survival. To provide a broad analogy, fire may be a vital component in many ecological systems, and may act to cleanse old plant foliage, but if fire raged permanently throughout living plant populations it would demolish them. Internal plant structure tends to delimit the spread of fire; for example, new, sap-filled foliage is more resistant to fire, whereas old dry foliage is susceptible. According to this view, the story of evolutionary advance is the story of the degree to which natural selection — that process which "burns" or "consumes" variation — can be tamed and redirected by indigenous structures or processes within the genome.

What happens if the genome does not effectively limit the forces of natural selection? In evolutionary history, one of the anomalies in neo-

Darwinian theory is that after more than three billion years of relative evolutionary stasis in the pre-Cambrian period, a great explosion of variation occurred in the Cambrian era, beginning between 500 and 600 million years ago, a process which has not ceased until the present day. If, according to neo-Darwinian theory, natural selection was occurring before, during and after that period, what event or events triggered this relatively sudden acceleration in evolutionary variation? It surely cannot be explained entirely by natural selection alone. One of the testable hypotheses that emerges from our analysis is that it was the evolving ability in the eukaryote genome to resist and control the variation-depleting processes of natural selection, and to introduce significant variation against the current of natural selection, which gave rise to the eruption of variant types in the Cambrian period.

In this critical area, the difference between neo-Darwinism and silent gene theory could not be more marked. In neo-Darwinian terms, the genome is subject to the processes of natural selection, which in turn drives evolution. In silent gene theory, natural selection constantly reduces variation, and the eukaryote genome enhances its survival by resisting, in numerous ways, the variation-depleting effects of natural selection. Complex, rapidly evolving multicellular life begins to develop only when the genome achieves these survival techniques.

As a means of elucidating the difference further, let us consider what would happen if the silent genes, and the other silent processes which control the inroads of natural selection in the genome, were "switched off". It is our postulate that the evolutionary system would undergo a slow but progressive meltdown. The variation-reducing effects of natural selection would proceed to move unchecked through the system. All complex species would eventually become "brittle" in their niches and in due course would become extinct as the result of environmental changes. With variation constantly being reduced by natural selection, the end result would be a few, simple, generalised species representing perhaps the last traces of animal and plant life.

From the perspective of silent gene theory, life in pre-Cambrian times tends to resemble those conditions which the theory predicts if Darwinian natural selection were the main evolutionary force: that is to say, conditions of minimal morphological variation, little significant development of complexity, and relative evolutionary stasis. This would be the case despite aeons of time, vast population numbers of simple, generalised organisms and a rapid rate of generational turnover — according to classical Darwinian theory, precisely the conditions which should generate fast evolution.

It is our contention that only when we begin to accept that the central survival function of the genome is to overcome the variation-depleting effects of natural selection are we able to evaluate its various characteristics beneath a single coherent light. Under this light, it is submitted, much of the structure and many of the processes of the genome appear to be revealed in a more comprehensible form.

New discoveries of silent gene regulation

Until now, we have attempted to generate a broad narrative in which the conventional assumptions of the theory of natural selection are challenged, and a set of radically different basic rules are proposed in their place.

The outlines of silent gene theory were first proposed in March 2000, and the majority of this particular work was written before 2002. Perhaps necessarily, in the absence of support from the biological community, it relies heavily on *a priori* reasoning. In terms of potential scientific usefulness, therefore, its claims must rest on a sufficient degree of explicitness to be testable.

Since 2002, however, there has been an increasing, though somewhat piecemeal, recognition of the potential importance of the silent areas of the genome. As a means of assessing the general plausibility of silent gene theory, it is useful perhaps to consider this recent research more closely.

It was suggested earlier, for example, that it is in the "interest" of silent genes to "co-opt" and control other genes, but so far this has been entirely hypothetical, and no precise mechanisms have been proposed. However, in what appears to be one of the most important developments in recent biological science, a number of researchers have begun to investigate this new frontier entirely independently of silent gene theory, and are unearthing what seems to be a remarkable series of corroborations.

Amongst such researchers, John S. Mattick and others have identified a new level of control and regulation in the silent genes of more complex species. In addition to driving evolutionary variation, the silent genes in the genomes of complex organisms encode (in Mattick's words) "a vast and hitherto hidden regulatory layer".

According to Mattick and other researchers, the system works through RNA rather than DNA. When DNA creates proteins, it uses RNA as a "messenger". But it appears that silent parts of the genome generate RNA which is not used to make protein but which is deployed instead to precisely regulate much of the genetic system. The implications of this rapidly expanding field of investigation are revolutionary. The research uncovers an alternative function of silent genes which appears to derive directly from the evolutionary functions ascribed to them in this paper.

Mattick, Taft, Pheasant and other researchers have investigated a number of peculiarities of silent genes, and have reached the conclusion that a significant fraction of the silent part of the genome is engaged in the regulation of coding genes. In a major paper called, appropriately, *A New Paradigm for Developmental Biology*[9], published in 2007 in the Journal of Experimental

Biology, Mattick has summarised the comprehensive evidence in favour of this view. His broad conclusion is that genes which code for protein construction play the major genetic role in relatively simple organisms like prokaryotes, but in more complex unicellular eukaryotes a more sophisticated method of gene regulation began to develop about a billion years ago.

RNA is far more compact than proteins, and is well suited to hold and transmit complex information. According to Mattick, the encoding of information in a complex network of RNA allows both greater complexity and more precision of genetic regulation. Mattick elucidates the RNA regulatory structure through analogy with sophisticated modern engineering products. According to this view, protein coding is not unlike building the physical structure of a modern aircraft — it generates the physical engineering of the organism. By contrast, the silent genes, coding for RNA rather than DNA, create a system similar to the digital systems controlling the functions of the aircraft.

Mattick stresses that the recent sequencing of genomes of a wide range of species strongly supports the thesis that these more sophisticated systems of RNA coding and regulation first began to appear in the genomes of certain single-celled eukaryotes about a billion years ago in the non-coding part of the eukaryote genome. According to Mattick, further evolution of regulatory frameworks within the non-coding genes in turn helped to pave the way for the Cambrian expansion into complex multicellular organisms.

When considered as part of wider evolutionary processes, Mattick's informational theory of silent genes does not contradict our theory that silent genes drive evolution, but instead reinforces it and intertwines with it. As Mattick argues, to evolve complex new variations is not enough — the systems must be administered and regulated, and this increasingly sophisticated regulation must be part of the evolving system.

Although Mattick's argument is lucid and convincing, the fact that the appearance of a new layer of regulatory silent genes may give rise to the possibility of a rapid development of a more sophisticated structure does not, in itself, explain *how* the evolution takes place. Some driver is needed, and we submit this driver is the fertility of variation created by the silent genes.

Such an hypothesis — that increasing sophisticated informational and gene regulation networks derive from silent genes — is in principle testable. If silent genes generate variation and drive evolution, we would expect to see informational functions evolve from pre-existing banks of silent genes. In due course, informational silent genes should exhibit generic lines of development from earlier silent genes. Since, at first sight, informational silent genes appeared amongst silent genes following the appearance of silent genes in significant numbers in the early eukaryotes, it seems this prediction has a good chance of being verified. Further research will continue to throw fresh light on the matter.

Mattick's informational theory appears to give further weight to the central role of silent genes in driving evolution. In essence, our thesis proposes that variation generated by silent genes creates, over time, more sophisticated informational genetic systems of regulation and control. This enhanced informational capability in turn gives rise to greater potentialities for variation, in a mutually reinforcing relationship which may provide a further insight into the Cambrian explosion of variety and complexity.

Meanwhile, the function of this chapter is not to explore in any depth the new frontier of informational theory in silent genes, or to survey the impressive and rapidly growing body of research which supports and extends it. Mattick and other leading lights of informational theory have already provided both arresting evidence and a cogent *rationale* which is available in the research literature to all who wish to study it in detail. Rather, our aim is to acknowledge the potential significance of informational and regulatory functions amongst the silent genes, to describe their outline and indicate their likely place within the broader predictive structure of evolution driven by silent genes.

The purpose of the remaining chapters is to rehearse the arguments in favour of a silent gene theory and to observe the broader structure of evolutionary theory from its radically altered perspectives. These differences are so pervasive when compared with orthodox evolutionary theory that perhaps the reader will forgive occasional reminders of the main features which constitute such a transformed landscape.

Initial derivations and predictions

Initial derivations and predictions from silent gene theory are likely to include the following:

The theory should stimulate intensive and thorough modelling of the processes of natural selection to find out whether natural selection creates variation by virtue of adaptation to a heterogeneous environment (neo-Darwinism) or whether natural selection actively reduces variation within all environments (silent gene theory). The neo-Darwinian models should include the synthetic view — that variation is produced by mutations amongst genes which code for characters. The silent gene models should take account of the effects of additional mutation along free pathways amongst the large preponderance of silent genes.

If active and direct association of a gene with a physical character in the phenotype is the precursor to the winnowing effects of natural selection, it is a prediction of silent gene theory that genes which directly code for a physical character should be on average less long-lived than genes which are not actively associated with physical characters. By contrast, "silence" in a gene should be broadly correlated with longevity.

A further clear prediction concerns the ratio of silent genes within the genome, such that the ratio of silent genes $R^s = G^s/G^t$, where G^s represents the number of silent genes and G^t the total number of genes. According to silent gene theory, there should be a positive correlation of high values of R^s (the ratio of silent genes in the genome) with high rates of evolution, adaptive success and species longevity.

Since one of the long-term consequences of increasing variation within a species may be speciation (the formation of new species), it is a further prediction of the theory that active streams of evolutionary development, in which there are high rates of subsequent speciation, should also be associated with initially high R^s values. To give one example, in the case of the rapid mammalian spread during and after the major dinosaur extinctions, the broad pattern we might predict is that the original mammalian ancestor existed for a long time in the shadow of the dinosaurs, confined to relatively few niches because other niches were already actively filled by dinosaurs. Suppose, however, that towards the end of the dinosaur period the gene pool of the mammalian ancestor exhibited high R^s values, suggesting high potential variability.

By comparison with mammals, it may be that the dinosaurs during their final period had lower R^s values. If our surmise is correct, the incidence of low R^s values would make many species of dinosaur "brittle" in their various niches, and potentially subject to wide-ranging extinction, not least when a major environmental perturbation occurred, such as a climate-changing asteroid impact or a major volcanic explosion with planetary fallout.

By contrast, large R^s values for the ancestral mammal would make it highly adaptable not only to its own changing environment but also, potentially speaking, to a variety of other environments. So might begin the great expansion of mammalian variation and speciation, driven by the high ratio of silent genes in the ancestral mammalian gene pool.

Empirical evidence is beginning to emerge which offers strong support to the thesis that high R^s values are related to rapid evolutionary development. In 2003 R. J. Taft and J. S. Mattick[10] analysed the ratio of non-coding to total genomic DNA for 85 sequenced species which ranged from prokaryotes to *Homo Sapiens*. They noted that the conventional assumption that the number of coding genes and species complexity would be positively correlated was found to be relatively weak. Quoting Taft and Mattick's summary:

> *Prior to the current genomic era it was assumed that the number of protein-coding genes that an organism made use of was a valid measure of its complexity. It is now clear, however, that major incongruities exist and that there is only a weak relationship between biological complexity and the number of protein-coding genes. For example,*

using the protein-coding gene number as a basis of evaluating biological complexity would make urochordates and insects less complex than nematodes, and humans less complex than rice.

What Taft and Mattick discovered instead was a powerful correlation between species complexity and the ratio of non-coding genes to overall DNA — exactly as predicted by silent gene theory. It should be emphasised that this finding is highly anomalous in terms of neo-Darwinian theory, which stresses the importance of natural selection acting upon the coding genes, and whose practitioners have described the silent genes as "junk". By contrast, the genomic data strongly suggest instead that the coding genes are peripheral to the evolutionary process and the silent or junk genes are central.

The 85 sequenced species analysed by Taft and Mattick spread across a wide range, from relatively simple prokaryotes to more complex unicellular eukaryotes to highly complex multicellular organisms. To quote from their paper, the ratio of silent genes to total genes "is generally contained within the bandwidth 0.05 – 0.24 for prokaryotes, but rises to 0.26 - 0.52 in unicellular eukaryotes, and to 0.62 - 0.985 for developmentally complex multicellular organisms." The highest ratio (0.985) was found in *Homo sapiens*. The statistical correlation between the ratio of silent genes and complexity of the species was remarkable, leading Taft and Mattick to the following tentative conclusion:

> *DNA previously regarded as genetically inert ... may be far more important to the evolution and fundamental repertoire of complex organisms than has been previously appreciated.*

More recently, further investigation of the ratios of silent genes to the total genes in the genome — for example that of S. Ahnert and colleagues[11], who analysed 37 sequenced eukaryote species — has underlined the broad correlation between the non-coding gene ratio and the complexity of various eukaryote species.

As we have mentioned above, Mattick, Taft, Pheasant, Ahnert and others are investigating another feature of silent genes than their capacity to drive evolution — their ability to control and supervise the behaviour of coding genes, largely through RNA molecules. This potentially vast area of research is still at an early stage. The outlines, though, are reasonably clear. If species complexity is proportional to the ratio of silent genes in the genome, it appears increasingly likely that silent genes are responsible not only for generating such complexity but also for ensuring orderly processes by which the organism regulates its own activities.

Mattick and others are currently investigating precisely how non-coding DNA can regulate genes and their expression. For example, Haussler[12] has put forward evidence that a region of largely silent

DNA called HAR1 appears to affect cortical development in humans during embryonic development.

Taking these matters into account, the notion of a complex panoply of non-coding genes fulfilling further functions, such as regulating the coding genes, is consistent with the view that silent genes are at the heart of evolution. In addition, one deduction from the increasing preponderance of silent genes in complex species is that those vast arrays of non-coding genes are likely to perform complex and multiple functions, in which different areas of silent DNA incline to specialise in different directions.

It is our postulate that this account of evolution, outlining the central importance of the silent gene ratio (R^s), adds depth and weight to the explanation of certain major developments in the evolutionary saga, and in the process provides an important new strategic gene ratio against which to assess such developments. Indeed, if the hypothesis that R^s values contain a key to successful adaptation is valid, reliance on the vague postulate of large-scale external events to explain such evolutionary phenomena as extinctions might no longer carry the same significance, not least because it begs the important question of why other species survived and expanded during the same and subsequent periods.

In the history of evolutionary theory, W. D. Hamilton's powerful and original insight in considering evolutionary development from the perspective of the individual gene has given rise to what has sometimes been called the second Darwinian revolution. There is little doubt that its theoretical and practical yields have been significant. Yet even now, after several decades of impressive development and application, it has not fully resolved a number of classical problems, concerning which silent gene theory may perhaps offer a third wave of attack.

We have already argued that the action of natural selection on the phenotype is counter-variation, and would appear to undermine the evolution of more complex life-forms. In terms of Hamiltonian gene-centred theory, if the survival strategy of the individual or gene is to increase the number of its progeny in succeeding generations, then it may be argued that smaller, simpler organisms multiply at a faster rate, and such organisms therefore appear more apposite vehicles for successful reproduction. However we regard the problem, the evolution of complex organisms remains highly anomalous within both the classical theory of natural selection and its Hamiltonian offshoot.

By contrast, it would seem that the evolution of increasing complexity can be tentatively specified using silent gene theory. One of the consequences of the continuous accumulation of silent genes — because silent genes last longer on average than genes which are directly associated with physical characteristics — is that such accumulation slowly but progressively increases the size, and therefore the potential complexity, of the genome. A larger and more complex genome — particularly if it contains a high proportion of silent genes — in turn is broadly capable of evolving increasingly complex organisms. In considering cause and effect, it is our thesis that the accumulation of silent genes is likely to have led the way towards increasing complexity, not vice versa. This thesis is also in principle testable across a wide range of evolutionary data.

If it is the case that the central mechanism in the evolution of complexity is the accumulation of silent genes, another striking feature of evolutionary development — sexual reproduction — may also perhaps be better understood. If silent genes accumulate in organisms which reproduce asexually, then the fate of those silent genes is directly linked to the genes which are actively associated with physical characteristics in that single line of descent. By contrast, in sexually reproducing species, silent genes, freed from direct attachment to a single line of coding genes, and constantly mixed with other genes through sexual reproduction, will last longer on average than coding genes, which are subject to natural selection acting on the phenotype. It is therefore in the "interest" of the silent genes to reproduce sexually in order to increase their own ratio in the genome.

How can the silent genes in the genome direct their host organism to spread themselves by means of sexual reproduction? Considering the problem at one remove, it may be that a critical point is reached when the silent genes attain a significant numerical dominance in the genome. In any given species, it then becomes in the survival interest of the majority of (silent) genes in one organism A to combine sexually with the genes in another organism B. Similarly, it is in the interest of the majority of (silent) genes in organism B to combine with the genes in A. There would thus appear to be strong mutual interests in favour of sexual recombination.

In principle certain consequences of this reasoning are testable. Sexually reproducing species should be associated with higher R^s values than asexually reproducing species. Species which utilise both asexual and sexual reproduction, or which incorporate both means of reproduction at different stages, should be characterised by intermediate R^s values.

Another potentially important area of future research concerns the genesis and history of individual genes. According to our hypothesis, a gene may be randomly replicated, following which it may subsist for a considerable time in a silent form. During this time it may be affected by a number of mutations. In due course it may be re-integrated into active use, either by internal mutation or through the effect of other genes.

In pursuing the details of this proposed life cycle, there are numerous subsidiary hypotheses which could also be investigated, including the degree to which modifiers and certain genes which affect or govern the performance of other genes may be responsible for activating or co-opting silent genes.

Because no previous evolutionary theory appears to have concentrated attention on this area, one of the consequences of silent gene theory may be to encourage further research towards discovering the precise means and stages by which genes may emerge from silence into active coding for proteins.

Associated areas may also deserve attention. For example, polygenic activity, in which a number of genes influence one character, is common within eukaryote species. Yet how precisely do polygenes emerge? Do they form as a loose aggregation, or do certain active genes "switch on" or absorb other genes, perhaps from the reservoir of silent genes?

It is the submission of this paper that the above account merely scratches the surface of a potentially fertile yield of possible predictions from the silent gene theory of evolution. In almost every aspect of evolutionary development, from current adaptation processes to deep evolutionary history, it is argued that silent gene theory offers both alternative explanations and a fertile range of differing, testable subsidiary hypotheses.

An outline of evolutionary history

Since many of the preceding arguments must appear surprising, if not extraordinary, to the great majority of us who have been taught evolution in terms of neo-Darwinian theory, perhaps our interests are best served at this stage if an attempt is made to outline tentatively an alternative "history" of evolution, threading together the various assumptions, assertions and subsidiary hypotheses outlined above. At the risk of repetition, it may be argued that the terrain of silent gene theory is so strange that the larger pattern of evolutionary history deserves to be considered afresh from its radical perspectives.

Because these are early days in silent gene theory, we are obliged to tell our story in a relatively crude, outline form. In addition, the detailed elements of that theory may require considerable readjustment when the account is compared more closely with detailed current and future data from the wide number of fields to which it refers. Nevertheless, we submit that it is a potentially more coherent story than the neo-Darwinian account, and that in a developed form it appears more capable of precise predictions. The key features of this alternative history of evolution appear to be as follows:

The first several billion years of the evolution of life, until the Cambrian period, produced little more than unicellular organisms. The era was effectively dominated by natural selection which, acting largely alone and uninhibited, constantly reduces variation. The result, we submit, was relative evolutionary stasis in morphological form. The early single-celled organisms which slowly emerged were mostly built on the model of the prokaryote cell. The feature of the prokaryote cell which most inhibits the evolution of complexity is that there is a direct relation between the majority of its genes and the phenotypic structures for which those genes code. It is our surmise that this type of cell cannot easily evolve complex, varied organisms largely because the huge potential number of "silent" pathways which exist in eukaryote genomes are not available. Prokaryote genes do of course mutate, but the variations generated by these mutations are likely to be swallowed up by the strong counter-variation of natural selection acting on the physical characters of the phenotype. Expressed alternatively, the genetic structure of the prokaryote cell places it directly at the mercy of the variation-reducing effects of natural selection.

Although the prokaryote cell has not developed beyond a relatively constrained single-celled body plan, sheer numbers and generational turnover can provide considerable variation within that body plan. This apparent fertility of variation around a constrained body plan may be assisted by processes outlined above, in which spore stages, by removing genes temporarily from coding functions for life processes, act as the facsimile of a reservoir of non-coding or silent genes, and in which geographical factors, such as shore-side rock pools, may generate separately evolving sub-populations.

Even relatively simple prokaryote cells contain a small number of silent genes. Random replication of silent genes over time gradually increased the ratio of silent genes in certain prokaryote populations (as we have seen, Taft and Mattick indicate a significant range of 5%-24% in the ratio of silent genes among modern prokaryotes[10]). Silent gene theory predicts that lines of prokaryotes with comparatively larger ratios of silent genes evolved more rapidly and exhibited greater capacity for adaptation. Slowly, a different type of cell emerged, in the form of the single-celled eukaryotes, with a nuclear membrane separating the genome from the rest of the cell and a significantly higher silent gene ratio (26% - 52% among modern unicellular eukaryotes). The higher ratio of silent genes in turn ensured the cell was more capable of "resisting" the variation-depleting effects of natural selection.

How, precisely, does such a cell resist? Firstly, those genes within that cell which are "silent" are immune, or at least more resistant, to the variation-reducing effects of natural selection acting on the phenotype. We submit that in the main these are not new genes, but copies of existing genes which are, for the time being, silent or non-coding. Some of these may be "switched on" in due course, either in the genes themselves or by mutations or processes in other genes which may "co-opt" them. During the time in which these genes are non-coding, they are capable of mutating through a number of further non-coding states before reaching a new functional configuration. A large number might continue in the silent stage to develop mutations which render them temporarily useless; these would then become "pseudogenes". But as we have attempted to demonstrate above, even "useless" pseudogenes, after one or a number of further mutations, may reach configurations or combinations which in turn may be utilised in new contexts.

Although all new mutations are more likely than not to be deleterious when they begin to code, the probability of generating effective new coding genes is greatly enhanced by mixing copies of genes, or components of genes, which have already provided effective coding genes, or parts of effective coding genes, for use in other contexts.

About a billion years ago, as Mattick and others have proposed, a proportion of the silent genes began to evolve in the direction of a sophisticated regulatory structure, based on RNA, which in turn could make more precise and detailed use of coding genes. Because this "silent" structure of evolution and regulation existed at one remove from natural selection acting on the phenotype, it was more capable of resisting the variation-reducing effects of natural selection, and of generating increasing amounts of variation.

Perhaps it is useful to digress briefly here, and pose the following question: what is the theoretical structure of a genome which is capable of resisting the variation-reducing effects of natural selection? It is our hypothesis that one of the means by which such a genome would overcome these effects *is to break, in various ways, the simple one-to-one link between a gene and a physical character in the phenotype*. When the one-to-one link exists, natural selection, if it selects against the physical character in question, reduces the ratio of individuals which have that character, in turn reducing the ratio of genes which code for that character within the population. Thus the one-to-one link is the means by which natural selection can act directly upon the genome and reduce gene variation in the coding genes.

How, then, may that direct link be broken? Through, for example, pleiotropy, where one gene affects several characters of the phenotype, which in turn makes that gene far more difficult to eliminate. There may be a selection pressure against one of the physical characters for which a gene codes, but perhaps another counterbalancing selection in favour of a second physical character for which it also codes.

Another means of breaking or at least complicating that direct link is through polygenes and modifiers, so that the pathway by which natural selection can eliminate a particular gene is further blurred by the complexity of gene interaction. Yet another means is though the structure of dominance and recessiveness. Newer (recessive) mutations can be protected from the full force of natural selection because they are silent in the heterozygous state and only exposed to natural selection in the homozygous state. (Recessiveness, in this sense, appears a remarkably effective and tenacious system for maintaining non-dominant alleles in the gene pool. If x is the fraction of a recessive gene in the population relative to population numbers, then the chances of a recessive homozygous conjunction for any such gene are x^2, and it becomes almost impossible to eliminate entirely a recessive allele as its numbers depreciate. The longer such recessive alleles linger, the greater the chance that either they will mutate into a fitter form, or that a change in the environment may tip the balance in their favour.)

Indeed, it is our submission that almost the entire panoply of processes and structures within the eukaryote genome can best be understood in terms of their contribution to the genome's ability to resist and overcome the variation-depleting effects of natural selection.

Once the eukaryote genome is capable of overcoming the variation-reducing effects of natural selection, and of creating variation *sui generis*, the potential for evolutionary development becomes effectively limitless, or infinite. The next stage is the creation of complex multicellular animals, which in turn gives rise to the Cambrian "explosion" of variation and development in complex multicellular organisms. Since the Cambrian explosion, evolution has proceeded at rapidly increasing speed until the present day. Recent revelation of the powerful correlation between the ratio of silent genes in the genome and complexity of the species provides a further corroboration of the theory that silent genes drive the evolutionary process.

A radical rearrangement of evolutionary forces

At first sight, such an account may appear teleological to a degree. In answer, we can only attempt to emphasise that the central element in the entire silent gene theory — and the keystone to all the above structures and processes — is the assumption that variation-reducing natural selection is life-threatening to a complex and varied population; that natural selection endangers the survival of the individuals within that population continuously, from one generation to the next; that it is likely to be a more powerful, more comprehensive and even more persistent threat to the existence of the individuals in any given population than, for example, the "local" and perhaps more nebulous longer-term threat of "changes" in the environment.

Indeed, pursuing this reasoning to its logical conclusion, the key to overcoming the variation-reducing effects of natural selection is to generate a spray or width of variation from generation to generation which may then become the raw material for any process of further adaptation.

By "solving" the central problem of generating variation, the eukaryote genome revolutionises the rate at which complex evolution can progress. One possible prediction derived from this account is that the various recognisable features of silence in eukaryote cells appeared before the early part of the Cambrian period, in the form of a clearly recognisable pattern of increasing complexity in the silent genes rather than the coding genes. Part of the newly evolving function of the silent genes was to control other genes, generating what Taft and Mattick have described as a further layer of gene regulation[10]. This further regulative layer, with its enhanced informational capacity, in turn permitted further possibilities of variation.

The theory suggests another fertile area of investigation. Did the emergence of silent genes precede the main evolution of the eukaryote genome, with its distinctive gene nucleus separated from the rest of the cell by a nuclear membrane, or was the emergence of silent genes part and parcel of that evolution? In silent gene theory, any evidence which suggests that the appearance of significant numbers of silent genes preceded the evolution of the characteristic structure of the eukaryote genome would add weight to the argument that silent genes drive evolution.

The story tentatively outlined above differs from the neo-Darwinian theory of selection in one vital structural sense. Neo-Darwinism proposes that the primary generative force in evolution is natural selection. Silent gene theory proposes that the evolution of complex life consists of the contention of two forces, the "positive" process of indigenous variation generated by the silent areas of the genome, and the "negative" or winnowing process of natural selection.

The entire silent gene argument depends strongly on this single base. If, for example, it is assumed that natural selection is a creative or generative process, then the evolving ability of the eukaryote genome to "resist" its variation-reducing effects becomes largely nonsensical, and all subsequent arguments take on a hollow, teleological character. In this sense, we submit, silent gene theory cannot easily be absorbed into the classical structure of neo-Darwinism. It is a separate theory of evolution, based on radically different principles and on a comprehensive reordering of the primary forces of evolutionary development.

By contrast, however, it is our claim that the great tradition of Darwinism can be subsumed, relatively easily and painlessly, within silent gene theory. This is because silent gene theory takes as given that natural selection exists, and that it constantly acts to reduce less fit variants in any population. If we were to summarise the relation between the evolutionary effects of silent genes and natural selection, we could say that silent genes propose, and natural selection disposes.

Under the aegis of silent gene theory, Darwinian natural selection continues to exert an extremely powerful but essentially secondary effect in evolution. One of its most significant effects is the high natural rate of extinction of species which natural selection, acting against variation, renders brittle in their environments. According to this view, natural selection exists and is powerful, but it acts in a somewhat different manner from that originally proposed by its great discoverer.

Our conclusion, taking the above into account, is that evolution is not so much the story of adaptation as the story of variation. The capacity to vary, not the capacity to adapt, more cogently defines and illuminates virtually every aspect of the evolutionary saga — from molecular genetics to the salient features of evolutionary history. Far from the silent genes being peripheral or parasitic or "junk", silent genes lie at the heart of the evolutionary process.

This "heart" consists of a quiet but immense engine of silent genes, which are wholly or partially protected from natural selection acting on the phenotype, and whose numbers slowly increase in the genome through differential selection. "Unseen" by natural selection, large numbers of mutations can accumulate in the silent genes which do not deleteriously affect the host organism. When finally switched on, exotic new variants are created and are then subject to natural selection. As with all new mutations, most of these new variants will be adverse but some will not, and these latter will drive the evolutionary process.

Species evolve at a rate proportional to the ratio of silent genes in their genomes. Part of the evidence for this slow explosion of indigenous variation created by the silent genes is that after approximately four billion years of evolution there exists all around us — not unlike the faint hiss from outer space which denotes an earlier primal explosion — an almost perfect correlation between the ratio of silent genes and species complexity.

Discussion

When a new theory is proposed, there is a tendency amongst the practitioners of a discipline to regard it as an exotic newcomer against the well-tried and the proven.

This is no doubt justifiable in most cases. Yet Kimura[13], for example, has pointed out that the empirical base of the prevailing modern synthesis was, and remains, extremely narrow. Even after it had been widely accepted following the persuasive work of Dobzhansky[14] and Mayr[15], amongst others, the so-called "direct evidence" of beneficial mutations of the coding genes appeared to consist of the almost totemic example of industrial melanism in moths and (later) increased DDT resistance in insects. Indeed, it would seem that a single precise example of the process by which an identifiable coding gene mutates into a new functioning gene of significantly improved survival value has yet to be found and directly identified.

The reasons for this now seem obvious. The evolution of one coding gene into another without non-coding or silent intermediate stages appears extremely difficult. More than 60 years after the modern synthesis was proposed and cemented by Dobzhansky and Mayr, amongst others, it could be argued, as Kimura as averred, that the empirical base of the modern synthesis remains largely mythic.

This is not to criticise the usefulness of the modern synthesis, which has proved remarkably successful as a theoretical umbrella for combining Darwinian natural selection and Mendelian genetics. It is merely to point out that in comparing the new synthesis with silent gene theory, the steadily accumulating evidence is that silent genes perform the driving role in the process of generating new variant coding genes.

In the pursuit of eventual acceptance, a theory must first be proposed before it can be proved or disproved. *A Silent Gene Theory of Evolution* is necessarily at an early and highly tentative stage. Notwithstanding, we submit that silent gene theory deserves to be aired and assessed over the longer term in order to thoroughly test its strengths and weaknesses. Since it was first outlined in March 2000, a significant amount of evidence has accrued to the effect that the "junk" or silent genes are far more active than had previously been thought. A number of these features — including the central importance of the silent genes in mutation and variation — were predicted at an early stage in the formation of the theory, using largely *a priori* arguments. The weight of evidence now increasingly supports the initial premises. The more recent discoveries that silent genes provide complex layers of genetic regulation through RNA further underscores the assumption that silent genes are central to the evolutionary process.

Other factors impinge, however, and one in particular is too important not to be mentioned here. We are currently at a cross-roads in genetics, with a rapidly developing capacity to sequence the genomes of an increasing number of species. More than ever before, we need powerful and fertile theories with which to approach the vast new bodies of evidence uncovered by these technical advances. Against this background, silent gene theory offers a radical new account of the origination of effective evolutionary mutation and the source of predominant variation in evolution. It is potentially important, we submit, that this theory should be available to researchers in the next phases of detailed enquiry.

One of the key features of silent gene theory is that it offers a new biography of functioning genes which is markedly different from existing neo-Darwinian theory. In making this biography as explicit as possible, let us briefly summarise the three main rival theories (the modern synthesis, the neutral theory, and silent gene theory) as potential explanations of the origin of variation:

1. The modern synthesis postulates that coding genes evolve along (necessarily highly delimited) pathways of functioning intermediate stages in the face of the variation-depleting effects of natural selection. As we have mentioned above, direct empirical evidence of the generation of effective new mutant genes in the coding genes remains remarkably thin.
2. Kimura's neutral theory of evolution, which makes no vital distinction between coding and non-coding genes, proposes that the genes evolve along a neutral axis by means of drift.

Those predictions which can be drawn from the neutral theory about the details of evolutionary development have been found to be largely contradicted by the evidence.

3. In contradistinction to both of the above, silent gene theory unites two wings of evolutionary theory (gene mutation and phenotypic modification) by proposing that silent genes generate the variation which is necessary before phenotypic adaptation can occur. It suggests that the ratio of silent genes in the genome is likely to increase over time because of an internal selection process in which silent genes last longer than coding genes. Using new empirical evidence, this prediction is directly testable by assessing the longevity of silent and coding genes. In this state of silence, it is proposed that genes are able to evolve along free pathways, passing through durable, but non-coding, intermediate stages. Crucially, in this evolving mode, because they do not code for proteins, they are not subject to the variation-reducing effects of natural selection on the phenotype. This results, we submit, in a vastly richer supply of new gene configurations. In due course, silent genes may emerge from silence through various random processes and begin to code, or alternatively, to regulate the expression of coding genes. These new variations, like all mutations, are likely to be deleterious, but amongst the vastly richer supply of new

variations from silent genes there should be new viable genes and genetic interactions.

Of the three theories, it may perhaps be argued that silent gene theory is already much closer to the observed facts than the theory of natural selection or neutral evolution. Neither natural selection acting on the phenotype nor the neutral theory of evolution explains the selective origin and continued function of the majority of the genetic material in the eukaryote genome which (at any one time) does not code. Silent gene theory not only deduces the existence of this huge body of genetic material from clear *a priori* reasoning, but predicts its salient characteristics.

Whereas the theory of natural selection explains the function of that part of the genome which codes, silent gene theory explains why both coding and non-coding genes are necessary to evolution, and in the process it correctly predicts the increasing ratio of non-coding genes in more complex species.

Perhaps the most important part of our debate for geneticists, however, is that silent gene theory specifies a new and explicit biography of genetic mutation, which can be summarised as follows:

coding genes ⇒ random duplication ⇒ prolonged silence ⇒ accumulated mutations/substitutions via durable intermediate stages ⇒ random shuffling/construction of new genes ⇒

precipitation from silence/switching on by other genes $\Rightarrow$ active expression as new functioning genes

At the time of writing, full and detailed sequencing of the human genome and an increasing number of other genomes has already been achieved. Yet rapid progress in the mapping of the genomes of various species is only the first stage in a long process of elucidation of gene function which will take decades and perhaps longer still to develop fully. With this background, what matters now is how to analyse those gene sequences, how to deduce the detailed history of their origination, and how the overall structure fits together within developing evolutionary theory. Silent gene theory — which is likely to evolve from its *a priori* beginnings as it is taken up increasingly by researchers — should play an increasingly active part in this process.

POSTSCRIPT

If the problems of finding a publisher among peer-reviewed journals have been frustrating, there have been other compensations in attempting to formulate the beginnings of a silent gene theory of evolution. One of these is the generous intellectual support I have received from scientists who have a sincere and objective interest in evolutionary theory. It seems to be a pattern amongst readers with scientific backgrounds that those who are not professional neo-Darwinists are nearly always sympathetic to silent gene theory, and appear happy to cite the theory's various advantages in explaining evolutionary processes when set against the theory of natural selection. Equally gratifyingly, the favourable response often appears to be swift and certain. On more than one occasion I have received such a response after one reading and by return of post.

Two leading scientists in particular have both encouraged my efforts and raised my morale.

Professor Donald Braben, a nuclear physicist by background, over a ten year period led a celebrated series of advanced research programmes at British Petroleum, supporting projects whose subject matter ranged from molecular chemistry to complex cell biology. His method was to look for precisely those scientists whose proposals were sufficiently original and iconoclastic that they were not likely to find conventional science funding, particularly where that funding was governed by peer review. As always, it was difficult to weed out the genuine cases from the eccentric or fanciful, but by no means impossible. Braben says that one of the important lessons in assessing new projects was to "un-learn" existing assessment procedures.

By actively ignoring the process of peer-review, which tends to exclude both eccentric and original theories, Braben's success rate, in terms of pure research and economic advantage, was nothing short of remarkable. It is surprising that his principles have not been more widely emulated elsewhere.

In 2001 Braben wrote a letter to the two leading international science research journals *Nature* and *Science*, in which he was supported by several other distinguished scientists, including the Nobel laureate Professor Sir Harry Kroto and the distinguished Harvard chemist Professor Dudley Herschbach. Their joint letter argued that the institutionalising of peer review across all fields of research science actively prevented genuinely original papers from being published. Perhaps not wholly unpredictably, both *Nature* and *Science* refused even to publish the letter.

In Braben's view, if Darwin or Einstein were writing now, the radicalism of their views, combined with their detachment from contemporary science and their willingness to speculate or theorise outside the existing frameworks, would result in both being rejected by the process of peer review. In our era of over-specialisation, Braben argues, science needs its dissidents more than ever.

In an article in the *Times* of February 26, 2001, by Anjana Ahuja, the failure of *Nature* and *Science* to respond to the warning issued by Professor Braben and colleagues was reported. In the course of this article, Professor Braben was interviewed, and once again made the point that whereas modern science is awash with data as a result of impressive, high profile empirical research programmes such as the Human Genome Project, the mere mapping of genes did not itself constitute a sufficient explanation of their presence. An additional element of original theory was required to explain both their origin and function.

In healthy science, Braben argued, innovative theory and new empirical research should operate in unison. It is characteristic of strong and expansive phases of scientific advance that explanatory theory and empirical research are like two climbers linked together, each alternatively using the gains of the other to press forward. However, in phases of development which were lacking in original theories, science was in danger of becoming merely an increasingly impressive but often empty accumulation of information. By a curious coincidence Braben cited, as perhaps the most glaring example of the manner in which original theory had lagged behind data, the very fact that while we are currently able to map the human genome in precise detail, the function of the vast majority of non-coding genetic material in the genome was still largely unknown.

I happened to chance upon the article through the intercession of a friend, who had set it by as one of those definitive essays which one occasionally retains for future reference. Aware of my own difficulties in finding a publisher amongst peer-reviewed journals, he sent a copy of the *Times* article to me. In April 2002 I sent my paper on silent gene theory to Professor Braben and received, by return of post, a letter of 26 April which began:

> *I must say now that you make an excellent prima facie case. I like your silent gene theory very much. The concept has a ring of truth, and I would love to talk to you about it.*

Since writing his letter, Professor Braben has been a constant source of encouragement and useful advice.

My second source of encouragement came a few months later, with a note from Professor Freeman Dyson. Freeman Dyson is a British-born physicist who, at the remarkably young age of 29, in 1953 became professor of physics at the Institute for Advanced Study at Princeton University. The IAS is widely considered to be one of the most prestigious homes for theoretical science in the world. It is devoted entirely to research, with a strong emphasis on theory, and carries few or no teaching duties. Instead its members pursue whatever line of theoretical investigation may interest them. Amongst its famous alumni were Albert Einstein and J. Robert Oppenheimer. As its professor of physics since 1953, Freeman Dyson has been at the heart of theoretical science for the last half of the twentieth century. After he retired from administrative duties and became professor emeritus in 1993, he devoted himself to broader topics, including evolutionary theory, in a series of lucid expository works aimed at interesting a wider audience in the problems of theoretical science.

With some trepidation, I sent on a copy of *A Silent Gene Theory of Evolution* and received, again by return of post, Dyson's considered reply, in which the second sentence ran:

> *I like your theory and think it has a good chance of being right.*

Having praised the structure of the theory, however, Dyson proceeded to chastise me gently for presenting silent gene theory as "anti-Darwinian". He kindly added, "Darwin would probably have liked your theory." Dyson then proceeded to make the case that Darwin himself was highly aware of the difficulties posed by variation:

> *Darwin knew well that natural selection does not explain variation, and that the true explanation of variation remained to be found.*

In his letter Dyson outlined a plausible and coherent case for believing that silent gene theory should be considered not as a rival to the theory of natural selection, but a candidate for providing that "true explanation" of the source of variation which was necessary to complete the structure of Darwin's theory. I do not object to this clear and intelligent interpretation. On the contrary, if it were the case that silent gene theory were able to supply a "true" theory of variation, I should consider myself fortunate.

This central element of Dyson's well-argued letter stands by itself, and requires no further comment. Instead, I should like to consider one other aspect of his argument. In the course of his letter, Dyson draws a clear distinction between, on the one hand, Charles Darwin, and, on the other, what he calls the "self-proclaimed Darwinians".

> *The usage of the word "Darwinian" to mean a narrow reliance on only one mechanism is wrong. By using the word in this narrow sense, you fell into the trap that the self-proclaimed "Darwinians" prepared for you.*

Dyson is, I believe, fully justified in drawing such a distinction between Charles Darwin, the great naturalist and discoverer of the principle of natural selection, and certain modern "Darwinians" who purport to speak in his name. But despite Dyson's eloquent argument I have continued, until now, to compare silent gene theory with the beliefs of "the self-proclaimed Darwinians".

My reasoning is as follows. Any remaining difference between Dyson and myself on this matter, I would submit, arises partly from our very different positions within the firmament of science. Dyson occupies a well-deserved place at the summit of honour and prestige in theoretical science. By contrast I am an outsider, a maverick novelist attempting to set out a broad hypothesis on evolution. In his Olympian position above the clouds, Dyson is thoroughly justified in ignoring the "self-proclaimed Darwinians". But from my own perspective, it is the same Darwinians who control the peer-reviewed journals on the subject of evolutionary theory, and whom publishers of non-fiction would naturally and automatically consult over general publication of my work. I am, in this sense, *forced* to address the Darwinians precisely because they stand between me and the public I wish to address. By dogmatically insisting that their doctrinal form of Darwinism is the one true theory of evolution, the school of modern Darwinism effectively compels any proposed alternative explanation to set itself in rivalry with their own beliefs.

There is, I would submit, another related reason for addressing the Darwinians, which runs parallel to the first. A part of me cannot help but speculate upon how silent gene theory is likely to evolve in its relation to the theory of natural selection over the longer term. To underpin the argument, I have cited earlier Darwin's own profound words "unless profitable variations do occur, natural selection can do nothing." It seems to me that because the creation of variation precedes adaptation, and effectively occurs before adaptation can begin to act, silent gene theory in due course may come to occupy the central position in evolutionary theory.

With these thoughts in mind, I sent a copy of Professor Dyson's letter to Professor Braben, who was as pleased as I was to receive evidence of Dyson's cautious but generous encouragement. Braben began:

> *Thank you for your letter, and Freeman Dyson's comments. You have done very well. Freeman Dyson has long been one of my heroes ...*

Braben then proceeded to articulate a tentative alternative view on the possible longer term relation between silent gene theory and the theory of natural selection:

> *Hierarchically speaking, variation is of greater significance than selection. I agree, therefore, that if silent gene theory were proved correct, it would be the more complete*

> *theory, as Einstein's is compared with Newton's. However, it will be a long struggle, as you know. Old habits die hard.*

In the long term Professor Dyson may be proved right, as he has on so many other theoretical issues, and silent gene theory may end up supporting the theory of natural selection. However, if only for the joy of argument, and in order to place my own views (and perhaps Professor Braben's too) in a potentially refutable form, I hope the reader will forgive me for adding the following points.

If the central thesis of silent gene theory is correct — that variation precedes selection, and adaptation occurs downstream of variation — then natural selection is a secondary, not a primary, force in evolution. In this sense, it does not matter where one begins, the result will be the same. We may start from Professor Dyson's contention that silent gene theory seems to provide us with an effective theory of variation within the edifice of natural selection. After a while, however, a strong theory of variation will tend to take precedence over a strong theory of adaptation merely because it will become increasingly evident that, as the two theories are compared — to quote Darwin again — without "profitable variations" natural selection "can do nothing". In other words, the weaknesses of the theory of natural selection as the primary evolutionary force will become more apparent the more closely we compare it with silent gene theory.

Natural selection is based on the belief that evolution proceeds through the modification of physical characters. This in turn leads to the notion of the centrality of the coding genes, and this in turn has resulted in the appellation of terms such as "junk" or "pseudogenes" to genes which appear to be inactive in this respect. The manifest inability of Darwinians until now to either predict or understand the function of non-coding genes is only one demonstration of the inadequacy of the theory of natural selection as the central theory of evolution.

As silent gene theory predicts, in practice we find there are few known examples of significant evolutionary mutation in the coding genes, and increasing evidence of major and significant mutation among the non-coding genes.

In terms of silent gene theory, not only is natural selection a secondary — though important — process in evolution, it is almost entirely entropic and destructive. Natural selection is not nature's creator but nature's executioner — the equivalent of an abattoir in which those organisms which are less well suited to survival are cleared away. As we have argued, therefore, natural selection is not the creative force which neo-Darwinians claim for it. On the contrary, if natural selection were allowed to proceed unchecked, it would tend to reduce variation in any species to the point where that species would become more "brittle" in its environment, and less able to adapt to large or extreme environmental changes. This in turn increases the chances of extinction. Far from generating greater complexity, the variation-reducing effects of natural selection are more likely to reduce or eliminate complex specialised systems with greater levels of specialisation and interdependence, and are less likely to have lethal consequences in more simple, rapidly-reproducing species.

If we were aliens looking down on a planet Earth in which there were no silent genes, and natural selection were the predominant mechanism, we would argue that the surface of our planet would look as arid and uninhabited from space as Mars, or the moon. The metaphor is a broad brush illustration, of course, because the earth has water, and an atmosphere which might indicate by its chemical content that micro-organisms exist or were once present. But there is convincing evidence that Mars also once had water, and perhaps an atmosphere too. Looking at this hypothetical Earth from outer space — an Earth without silent genes — we might deduce that there could be tiny micro-organisms down there, like the prokaryotes (as there might indeed be such organisms somewhere on Mars, or on other planetary bodies), but no clear or visible sign of life seen from space. In the real world, by contrast, the abundant life that we actually see with our own eyes (and which we could also see with powerful telescopes from outer space) consists entirely of eukaryotes. In turn all eukaryotes have very large numbers of silent genes. Silent genes, not natural selection, give the earth its visible mantle of varied, complex, multicellular life.

I have mentioned above that the view that silent gene theory supplies the missing theory of variation in Darwin's structure, thus shoring up the main picture of evolution by natural selection, is a logical and cogent one. But there is another reason why I believe that silent gene theory may become the leading evolutionary theory in due course. When the two grand theories of evolution are placed alongside one another, and are allowed to evolve quite independently of my own views or those of my distinguished scientific mentors, I predict that over the next few decades silent gene theory will start to open up increasingly important areas of new research, and will come to displace the theory of natural selection in practical terms as the richer and more precise research paradigm.

If this were so, and taking into account Professor Dyson's important differentiation between Darwin and his modern supporters, where would this leave Charles Darwin, the historical figure, as opposed to the "self-proclaimed Darwinians" who claim to speak in his name?

For what it is worth, I believe Darwin himself will remain at the summit of evolutionary theory, as its founder and greatest theorist, though for rather different reasons than those which are currently put forward by the "Darwinians".

If it is the case that Darwin's original theory of evolution by natural selection of individuals will in due course be perceived as increasingly flawed — at least when proposed as the predominant force in evolutionary processes — this is not the case with certain other forms of naturally occurring selective process. There are other types of "natural" selection which emerge, for example, from silent gene theory, and which appear critical to the process of creating variation.

One of these selective processes is the proposed intra-genomic selection which favours silent genes over coding genes. Such a selection acts in a manner that could not possibly be perceived by a nineteenth-century naturalist who, through the circumstances of his time and place, knew nothing of genes. Even so, because it is a natural selective process, it may also be described, properly speaking, as a particular form of the natural selective process which was first expounded by Darwin.

Speaking for myself, I will continue to respect and honour Darwin's magnificent achievements and contributions to our understanding, particularly when set against the prevailing views of his era and the science of his time. The mere fact that we continue fiercely to debate that contribution in the twenty-first century is testament to his profound influence. I hope that this short work, published in the year of the 200th anniversary of his birth and the 150th anniversary of the publication of *The Origin of Species,* makes its own small contribution to celebrating that achievement.

In conclusion, the reputation of Charles Darwin — the great naturalist and founding theorist of the principle of natural selection — is likely to survive and prosper under circumstances in which silent gene theory is permitted to challenge the prevailing orthodoxy. But "Darwinism" — the dogmatic belief that the natural selection of individuals is the overwhelming and predominant driver of evolution — may not so easily survive silent gene theory.

The continuing importance of adaptation

In another sense, however, Freeman Dyson is likely to be right in his view that silent gene theory may complement, rather than replace, Darwin's theory of natural selection. If silent gene theory may in due course displace natural selection as the primary explanation of evolution, it does not displace what I conceive to be the most useful and important working element of our Darwinian inheritance — the widespread application of the theory of natural selection to the form and function of living organisms. To give only one brief example, the broad assertion that structures have evolved to perform certain survival functions still seems to me self-evident. Thus — to give only one example — evaluating the detailed structure of wings in terms of their perceived function of enabling an organism to fly appears entirely reasonable. The detailed study of adaptive evolution acting on structures and behaviours will remain largely untouched by the notion of silent genes.

Although I believe, in broad evolutionary terms, that adaptation is an important but secondary process which occurs downstream of variation, the argument that the organism evolves to suit its environment remains fully intact, and provides a consistently useful and informative means of considering detailed physical evolution. Furthermore, since the vast majority of practical "Darwinian" analysis is directly concerned with the process and mechanism of adaptation, there are almost no existing "Darwinian" works which strike me as incorrect in any significant manner, or which require to be comprehensively overhauled. In this sense at least, silent gene theory is likely to be a velvet revolution.

"Darwinism" only becomes dubious, or begins to appear unreasonable, when it asserts the increasingly unfounded metaphysical assumption that natural selection drives evolution, or is the primary force in evolution. This is particularly the case when it appears to go further still, and insists (as Richard Dawkins has argued in a recent Channel 4 television series on Darwin) that evolution by natural selection is a profound truth which is now so well established that it should not be questioned.

The next several decades will be a time when I suspect that the theory of natural selection and silent gene theory will contend at almost every level to explain the key processes of evolution. During that time, I submit, professional biologists would be wise not to assert too dogmatically that natural selection drives evolution before the new genetic evidence for either theory has been properly gathered and assessed. Until then, it would be better still if the theory that natural selection drives evolution were regarded as a working hypothesis and not wielded against opponents (or open-minded researchers) as a quasi-religious article of faith. Speaking for myself, as someone who believes all scientific knowledge is tentative, including silent gene theory itself, the removal of the theory that natural selection drives evolution from its current position as an absolute and unassailable dogma would be only one of various benign consequences of silent gene theory.

In this same spirit of hypothesis, I submit that the likely realignment of forces will be as follows. Natural selection in due course will be stripped of its metaphysical penumbra, and its quasi-religious aspects. It will be seen instead as a process which plays a strong part in evolution, amongst other evolutionary processes. It will continue to be treated respectfully and appropriately as a powerful working theory. Meanwhile its contribution to the broader scheme of evolution will slowly be re-evaluated. By contrast, its remaining practical core — the application of natural adaptation — will continue to lie at the heart of much continued useful investigation into evolutionary processes.

As events take their course, I suspect the relation between the two grand theories of evolution — natural selection and silent gene theory — will itself evolve much as Professor Braben suggests. If silent gene theory continues to prove accurate in its predictions, opening up new areas of research and offering new insights into traditional problems, increasingly it will be considered the more complete theory, "as Einstein's theory is considered to be the more complete relative to Newton." But it also seems to me that Braben's metaphor is richer than a simple comparison, and if the reader will forgive me a brief digression, I will outline some aspects of this evolving relationship a little more fully.

As someone whose occupation as a writer of sorts ensures that he spends a considerable time in a sedentary position at the word-processor, I happen to enjoy sea-fishing as a welcome break from routine. This entails going out in the elements, often at dawn or dusk, with a lightweight spinning rod, and tramping the seashore — in my case the South Hampshire coast. From my own perspective, the combination of beautiful surroundings and the sheer physical pleasure of throwing a spinner out to sea is sufficient reward in itself, without catching any fish. If I do happen to catch a mackerel, sea-bass or garfish which I can cook and eat, it is a bonus. The point I am leading towards is this. The natural environment which I inhabit as a fisherman is essentially Newtonian. It consists of the moon's effect on the tides which swirl around our local shores, the complex eddies and counter-eddies of water along the Solent shore, and other features of local weather, such as offshore or onshore winds, which are well explained by Newton's laws. Speaking as an amateur fisherman, I do not need Einstein to explain these dynamic features, even though I might admit, in an abstract sense, that Einstein's theories provide a more "complete" explanation of the structure of the universe.

The same is broadly true of the theory of natural selection. A vast amount of practical and useful biology can be carried out under the aegis of natural adaptation without reference to silent genes. But in another sense, the relation between silent gene theory and Darwinian natural selection is likely to be more complex than that of two entirely separate spheres of influence. Even as a fisherman, it would be a mistake to assert that we all live happily in a "sealed" Newtonian universe, and that Einstein's theories are only of interest to theorists. Einstein's theories impinge upon us in other ways. We have a right, for example, to be concerned about the fallout from nuclear testing, or possible pollution from nuclear power stations, and their long-term environmental effects on the sea and the creatures within it.

Similarly, I suspect that at certain points silent gene theory will impinge upon the practice of biology. It will strongly shape the way we look at the genome. The "silent" parts of the genome which until recently have been regarded as "junk" increasingly will be perceived as the powerhouse of mutation and variation, and thus of evolution itself. In refining our understanding of the evolutionary process biologists will begin to hunt down and clarify complex chains of cause and effect which until now have not greatly interested them. Provoked or perhaps inspired by silent gene theory, researchers will study far more closely how silent genes emerged, and to what extent the history of that emergence supports or disproves the theory. Perhaps one of the most promising results for silent gene theory would be evidence that silent genes first emerged amongst prokaryotes, and that the eukaryote genome subsequently arose from those prokaryotes where silent genes accumulated in their highest concentrations. Against this, the least promising result would be evidence that the full eukaryote genome emerged from prokaryotes before silent genes were present in large numbers. Advocates of silent gene theory would then be forced to explain how a complex structure like the eukaryote genome emerged without the impetus of silent genes. And so the process of dialogue between the two grand theories of evolution would continue.

In a wider sense, this is one of the attractions of silent gene theory. It "charges" the whole evolutionary debate once more by generating powerful alternative explanations of almost every salient feature or process in evolution. In addition, it poses questions which by their nature can be answered increasingly through sophisticated modern methods using the vast banks of computer data that are accruing on the genomes of a rapidly expanding bank of sequenced species. In this sense, too, silent gene theory promises to provide a theory which is wide-ranging and powerful enough to interrogate newer, more detailed and often highly recondite aspects of the evolutionary story, to improve our understanding and so, eventually, to put our increasing accumulation of data to more active and fertile use.

Scientific convention

How might this occur in practice? I admit that one of the oddities of this short book is that it is based to a greater or lesser extent on *a priori* reasoning (a point that Professor Braben makes in his lucid foreword). This, I suspect, is both a potential advantage and a disadvantage. One advantage of using *a priori* reasoning is that it connects the sometimes abstruse world of evolutionary theory with a common sense step-by-step approach which is more amenable to the general reader. A disadvantage is that it is not the normal form of the scientific paper, which generally attempts to elucidate new research data and keeps speculation to a minimum.

The reaction of biologists to *A Silent Gene Theory of Evolution* has shifted somewhat during the last nine years or so since I first put forward the outline of the idea to my former tutor John Maynard Smith in March 2000. The majority of *A Silent Gene Theory of Evolution* was composed between 2000 and 2002. Whereas in its early stages it was regarded by the biological community with incredulity, in more recent times, as evidence of the active contribution of silent genes has grown, I have found myself increasingly criticised by members of that same community for not using the rapidly developing research data on silent genes as the chief material of my thesis — in other words, for not conforming to the more accepted pattern of a scientific paper.

This is a large subject, and perhaps deserving of a more detailed consideration. But meanwhile I should like to remind these same critics that I am constrained to write from my own perspective. As noted in my introduction, I am not a professional scientist, I do not inhabit a scientific community, and I am somewhat detached from the details of the most recent research data. Taking this into account, I have always accepted that the work of relating silent gene theory in detail to the developing research data will be for others far more scientifically qualified and capable than I am.

In my defence, it seems to me that the primary function of *A Silent Gene Theory of Evolution* is to set out a broad description of the hypothesis that evolution is driven by variation rather than adaptation. The facts are that the theory developed through the use of *a priori* reasoning rather than by developments in genetic research, and in that sense at least I am being true to its genesis in casting this short book as largely an exercise in *a priori* argument. In the process, I have done my best to develop a silent gene theory of evolution in the form of a reasonably coherent story or narrative. In addition, in order to conform with my own Popperian beliefs that science is characterised by its capacity to be falsified, wherever possible I have attempted to set out the arguments in a potentially refutable form. My hope is that any subsequent dialogue between the two grand theories of evolution will continue to develop largely through extrapolation of the two theories to the accruing evidence, and through careful comparison of predictive capability.

As part of this argument, I submit that at this early stage the chief benefit of this approach is that as a concept silent gene theory is clearly and strikingly different from the current evolutionary model. Whereas the theory that natural selection drives evolution stresses the primacy of adaptation of the phenotype, silent gene theory suggests that genetic variation, and not adaptation, drives evolution. Where natural selection asserts that the external processes of selection are paramount, silent gene theory proposes that the causes of variation are largely indigenous to the genome. Where natural selection asserts that adaptation creates new variants and species, silent gene theory proposes that the process of natural selection constantly reduces variation to an optimum type, and is therefore negative, destructive, and entropic. These differences could hardly be starker, and are just as stark when we consider the respective functions of the genes themselves. Where the theory of natural selection asserts the centrality of the coding genes (as a consequence of which its advocates have traditionally designated non-coding genes as "junk") silent gene theory posits the centrality of the non-coding or "silent" genes both in generating mutational variation and in driving evolution.

Rather than interpreting the latest research data in detail, my chief objective has been to outline a radically different evolutionary theory which reinterprets the chief features of the evolutionary process. I have attempted to do so in a form which brings greater coherence to the salient features of the larger evolutionary story. To the degree that I have managed to fulfil such an objective, my purpose will have been achieved.

Some further implications of silent gene theory

How, finally, may we expect silent gene theory itself to evolve over the long term, not merely relative to Darwinism, but as a theoretical construct in its own right? And, more crucially, is this a question which, at this early stage, could even be asked with any degree of usefulness?

Technical matters aside, it is difficult to predict how theories develop, and what may flow from them. In physics, for example, wave and particle theory were each at various times predominant. Each was at different stages buoyed by new empirical data which seemed to favour first one side and then the other. Beyond this continuing dialogue with emerging data, history demonstrates that scientific theories often have effects which cannot be calculated by their initial propagators. Perhaps the most notorious example occurred when the somewhat reclusive and deeply humanitarian Einstein, calculating the energy in matter, set down the formula $E = MC^2$, only to find in due course that one of the unexpected consequences of his formulation was that the energy contained in matter could give rise to the atom bomb.

By the same token, how could the equally gentle and benign Charles Darwin have predicted that his theory of natural selection would be used at different times, and by ideologues of various hue, to justify movements such as social Darwinism, eugenics, Marxist class war, the more predatory forms of capitalism, even the active components of "survival of the fittest" in the ideologies of Nazism and fascism?

Like natural selection itself, silent gene theory may begin in due course to throw its own wider philosophical shadow. If it does, one thing at least is certain. It is a shadow which I, as one of its initial proponents, will not be able to control, even if I wished to do so. Perhaps, too, because I have struggled to formulate the initial outline of silent gene theory over the last few years, I have been too close to the concept to recognise certain aspects which may have become more obvious and interesting to those who have been kind enough to read my evolving typescripts. It is through their own insights, rather than my own, that we may begin to discern the faint hint of what this shadow might be.

To give an example, it has been pointed out to me on more than one occasion by readers that the suggestion that evolution is driven not by the coding genes, but by the large majority of silent or non-coding genes, has a number of fascinating implications of a philosophical and perhaps even a political nature. According to these commentators, it is as if evolution has generated, in the silent genes, a domain of freedom to mutate and evolve outside the regulation and conformity imposed on the coding genes by natural selection acting on the phenotype.

Unlike the Darwinian theory of natural selection, which suggests that fierce and relentless competition between individual organisms is the engine of evolutionary development, these readers have suggested that silent gene theory implies that the precondition of major evolutionary advance amongst the majority of genes is a state approaching a liberal and almost anarchistic freedom to evolve freely. Perhaps the interpolation of this new and radically different theory will also help a wider public to understand and take to heart the idea of evolution by natural processes. Meanwhile, the existing dogma that evolution is driven by natural selection, and that it behoves those of us who believe in evolution to acknowledge and to even worship the principle of elimination of the less fit remains — perhaps understandably — repugnant to many.

At various stages in my adult life, as recorded in the introduction to this essay, I have experienced the constraining influence of what Professor Dyson calls "the self-proclaimed Darwinians". Perhaps, against such a background, I might be permitted to extract a small degree of comfort from the fact that, if silent gene theory seems to suggest anything of a philosophical nature, it is that such doctrinal certainty of view appears inimical to the complex enterprise of learning about the world. Instead it would seem that — in science as in evolution — significant advance is difficult, if not impossible, without a wider and preferably predominant state of freedom.

REFERENCES

[1] Darwin, C., The Origin of Species, originally published by John Murray, (1859), Penguin edition, p 132, (1968)
[2] Hamilton, W. D., The genetical evolution of social behaviour, *Journal of Theoretical Biology,* **7,** 1 (1964)
[3] Kimura, M., Process leading to quasi-fixation of genes in natural populations due to random fluctuation of selection intensities. *Genetics* **39**: 280 (1954)
[4] Kimura, M., Evolutionary rate at the molecular level, *Nature,* **217** 624 (1968)
[5] Kimura, M., Population Genetics, Molecular Evolution and the Neutral Theory, 562, University of Chicago Press, (1994)
[6] Maynard Smith, J., Natural selection and the concept of a protein space, *Nature* **225:** 563 (1970)
[7] Ladle, R., Parasites and Sex: Catching the Red Queen. *Trends in Ecology and Evolution,* **7**, 405 (1992)
[8] Ridley, M., The Red Queen, Viking, 82 (1993)
[9] Mattick, J., A new paradigm for developmental biology, Journal of Experimental Biology, 210, 1526 (2007)
[10] Taft, R. and Mattick, R., Increasing biological complexity is positively correlated with the relative genome-wide expansion of non-protein-coding sequences, *genomebiology.com/ 2003/5/1/P1* [2003]
[11] Ahnert, S., How much non-coding DNA do eukaryotes require? *Journal of Theoretical biology,* **252**, *587* [2008]
[12] Haussler, D, *et al,* An RNA gene expressed during cortical development evolved rapidly in humans, *Nature,* **443**, 167-172 [2006]
[13] Kimura, M., How genes evolve: A population geneticist's view. *Annales de Génétique* **19,** 153 (1976)
[14] Dobzhansky, T., Genetics and the Origin of Species, Columbia University Press, New York (1937)
[15] Mayr, E., Systematics and the Origin of Species, Harvard University Press, (1942)